Indigenous Knowledge, Ecology, and Evolutionary Biology

Indigenous Peoples and Politics

Franke Wilmer, *General Editor*

Collective Rights of Indigenous Peoples
Identity-Based Movement of Plain Indigenous in Taiwan
Jolan Hsieh

The State and Indigenous Movements
Keri E. Iyall Smith

Speaking with Authority
The Emergence of the Vocabulary of First Nations' Self Government
Michael W. Posluns

Media and Ethnic Identity
Hopi Views on Media, Identity, and Communication
Ritva Levo-Henriksson

The State, Removal and Indigenous Peoples in the United States and Mexico, 1620–2000
Claudia B. Haake

Indigeneity in the Courtroom
Law, Culture, and the Production of Difference in North American Courts
Jennifer A. Hamilton

Politics and Aesthetics in Contemporary Native American Literature
Across Every Border
Matthew Herman

Indigenous Knowledge, Ecology, and Evolutionary Biology
Raymond Pierotti

Indigenous Knowledge, Ecology, and Evolutionary Biology

Raymond Pierotti

Routledge
Taylor & Francis Group
New York London

First published 2011
by Routledge
711 Third Avenue, New York, NY 10017

Simultaneously published in the UK
by Routledge
2 Park Square, Milton Park, Abingdon, Oxon OX14 4RN

Routledge is an imprint of the Taylor & Francis Group, an informa business

Typeset in Sabon by IBT Global.

Library of Congress Cataloging in Publication Data

Pierotti, Raymond John.
Indigenous knowledge, ecology, and evolutionary biology / by Raymond Pierotti.
p. cm.(Indigenous peoples and politics)
Includes bibliographical references and index.
1. Indigenous peoples—Ecology. 2. Traditional ecological knowledge.
3. Ethnoscience. 4. Evolution (Biology) I. Title.
GF50.P54 2010
304.2—dc22
2010002845

ISBN13: 978-0-415-87924-8 (hbk)
ISBN13: 978-0-203-84711-4 (ebk)
ISBN13: 978-0-415-51778-2 (pbk)

"[Crows and Ravens] have profoundly affected people all over the planet at every step of human evolution." (Marzluff and Angell 2005: 2)

On their way to their own homeland [The Apache] encountered a large Raven. . .[who] spoke to them: "It is our way to share what we have with others. Many times we invited you to come with us and you ignored our invitation. Our children played among you. . .They had come to invite you and wanted only that you should follow them. You would not do that. There will come a time. . .when our peoples will not speak directly to one another. ***We will have different languages, different homes, but we will always experience hunger****. That is the way of beings on this Earth. The next time our children come among you, do not ignore them. They have come to invite you to a feast." Violet Cordova (2007: 13, emphasis added)*

Descartes' philosophizing begins with the sentence, "I think, therefore I am." With this miserable, arbitrarily chosen beginning, it finds itself irrevocably committed on the road to the abstract. It never finds the door to ethics and it is caught like a prisoner in a dead world- and life-view. True philosophy must proceed from the most immediate and comprehensive fact of consciousness"I am life that wills to live in the midst of life that wills to live." This is not a subtly reasoned dogma. Day by day, hour by hour, I move in it. In every moment of reflection it stands before me anew—A vital world- and life-view that sees into all facts of being bursts continuously forth from it as from never-withering roots. The mysticism of ethical communion with being grows out of it." Albert Schweitzer (1950)

Contents

Figures

Acknowledgments

Many people have contributed to the ideas that I describe in this book as well as the manner in which I perceive the world. My mother and my maternal grandmother were the first people who showed me how to think in terms of connections to and relationships with the nonhuman world. I dedicate this work to their memories. My father and brothers were sometimes my companions as I explored the nonhuman world. We were regularly accompanied by a number of four-footed and winged companions of various species, who taught me about different perceptions and ways to explore the world that were beyond my ability to perceive without their aid.

My wife, Dr. Cynthia Annett, has provided invaluable help and work in editing and discussion of every chapter in this volume. Her solid grounding in the Western philosophical and scientific tradition has made her an ideal colleague for such efforts.

Over the years I have interacted with a number of Indigenous colleagues who have told me stories or discussed with me the manifold ways in which Indigenous people perceive and deal with the natural world, including Thom Alcoze (Cherokee), Joseph Bruchac (Abenaki), Vine Deloria Jr. (Lakota), Wilfred Denetclaw (Dine), Robin Kimmerer (Pottawatomie), Brent and Lori Learned (Arapaho), Henry Lickers (Seneca), Henrietta Mann (Northern Cheyenne), James Peshlakai (Dine), Enrique Salmon (Tarahumara), Dean Smith (Mohawk), Earl Swift Hawk (Lakota), Joseph Brophy Toledo (Jemez Pueblo), and Albert Whitehat (Lakota). Cliff Poodry (Seneca) allowed me to realize that it was possible for an Indigenous person to be a university biology professor. Joseph Marshall III (Lakota), the late Louis Owens (Choctaw/Cherokee), and Bill Penn (Nez Perce) have provided examples of how to write effectively and respectfully about Indigenous experience and the relationship to the nonhuman world. Cornell Pewewardy (Kiowa/Comanche) showed me that song can be used to communicate effectively across the barrier between species and that traditional songs can be used to comfort our nonhuman relatives.

A few individuals have worked closely with me and have read or commented on my writing and presentations in detail over the years. Greg Cajete (Santa Clara Pueblo) is a pioneer at demonstrating how Western science and traditional Indigenous beliefs share themes and knowledge and blazed the path that I have followed. Anne Calhoun (Eastern Cherokee) provided shelter, home-cooked meals, and invaluable advice. Donald Fixico (Creek/Seminole) was instrumental in developing the Indigenous Nations Studies Program at the University of Kansas. George Godfrey (Pottawatomie) worked with me at Haskell Indian Nations University on Indigenous science. Ronald Trosper (Salish) and Octaviana Trujillo (Yaqui) were my generous hosts during a sabbatical at Northern Arizona University during which I prepared the initial draft of this work. Francis Waukazoo (Lakota) has been a close and supportive friend over many years. Daniel Wildcat (Euchee) is a valued colleague with whom I worked out how to explain concepts from Indigenous and Western tradition. Michael Yellow Bird (Sahnish-Hidatsa) stood beside me over the years as we worked to convince the University of Kansas about the importance of Indigenous scholarship.

A number of non-Indigenous colleagues have taught me and shared ideas about Indigenous cultures and their philosophy and knowledge. To begin, I thank H. Gray Merriam, who taught my first course in ecology, thus allowing me to see that it is possible to have a science built on connections and relationships. Eugene Anderson and his marvelous book *Ecologies of the Heart* (1996) often provided inspiration. M. Kat Anderson showed me how different Indigenous people are from European environmentalists in the way they see the world. Russel Barsh and his writing have been a constant source of ideas. Douglas Causey invited me to participate in National Science Foundation supported meetings on Indigenous knowledge and provided support for a symposium I organized on Native American approaches to science. The late Bernard (Bud) Hirsch was a valued colleague in Kansas with whom I often debated the themes in Indigenous writing. Kerry Hughes provided encouragement and ideas. Henry Huntington showed me his work on Belukha, which provided a crucial early example that stimulated my thinking. Kelly Kindscher always reminds me that knowledge of plants is as important as knowledge about animals. Gary Nabhan welcomed me and taught a course with me at Northern Arizona University. He also graciously showed me his work and introduced me to the Seri people of Northwestern Mexico. Donald Worster is a colleague whose works have provided insight into the ways that different peoples perceive the environment. Finally, I have benefited from discussions and shared experiences in both Eurasia and America with my colleagues from the Russian Altai Republic: Yuri Tabakaev, Nicolai Malkov, the late Victor Lukyenenko, David Khaydarov, and Yulia Mechenikova.

My Indigenous students have been a constant source of inspiration and have taught me more than I ever taught them. They include Dustina Abrahamson (Shoshone), Temashio Anderson (Diné/Pomo), M. C. Baldwin (Diné), Nasbah Ben (Diné), Michael Benedict (Mohawk), Brenda Brandon (Seneca), Terry Brockie (Gros Ventre), Dana Carter (Chickasaw), Myron Dewey (Agui Diccutta band of Walker River Paiute), Cecelia

Flores (Alabama-Coushatta), Regina (Usti) Grass (Cherokee), Stan Holder (Lakota/Wichita), Katherine Humphrey (Cherokee), Catherine LaDoux (MicMac), Allen Long (Navajo), Heidi Mehl (Cherokee), John Ortley (Lakota), Cora Peshlakai (Diné), Jodi Potts (Gwich'in), Adrian Ricehill (Menominee), Jennifer Smith (Comanche), and last but certainly not least Corey Welch (Northern Cheyenne), who served as sounding board and discussant in many a situation.

This manuscript benefited from a full reading by and the extensive comments of Dorothee Schreiber, Charles Menzies (Gitxaala Tsimshian), and Loretta Jackson-Kelly (Hualapi). I also thank Ronald Trosper (Salish) and an anonymous reviewer for comments on earlier drafts of this MS.

Significantly shorter versions of Chapter 2, Chapter 7, Chapter 10 and Chapter 11 were published previously as:

Chapter 2: Pierotti, R. 2005. Communities as social and ecological entities. in *Being Native: Native Being*. Southeast Oklahoma State University, Durant, OK.

Chapter 7: Pierotti, R. 2002. Connected to the Land: Nature and Spirit in the Novels of Louis Owens. Indigenous Nations Studies Journal 3: 77–94.

Chapter 10: Wildcat, D. and R. Pierotti. 2000. Finding the indigenous in indigenous studies. *Indigenous Nations Studies Journal* 1:61–70.

Chapter 11: Pierotti, R. and D. Wildcat. 2000. Traditional ecological knowledge: the third alternative. *Ecological Applications* 10:1333–1340.

All new versions are greatly expanded and incorporate new material and ideas.

I would also like to thank the hundreds of undergraduates from both the University of Kansas and Haskell Indian Nations University who have participated in my classes—Native and Western Views of Nature, Natural Resource Management from an Indigenous Perspective, and Environmental Justice—over the last fifteen years. On two separate occasions (2002, 2006), I was privileged to participate in the course on ethnobiology at Northern Arizona University and had wonderful interactions with the students in those classes. Many other colleagues from a wide variety of cultures throughout the United States and Canada have shared their ideas and thoughts with me.

Finally, I developed and prepared this work with the generous support of the National Science Foundation through the grants NSF-DEB 93–17582, DEB 02–03404 and NSF CCLI 0717828.

Introduction

The Enlightenment was intended to be an intellectual movement that sought to emancipate human society from what it regarded as the triple tyranny of despotism, bigotry, and superstition (Worster 1994; Coates 1998). Although these seem to be noble ideals, the real trick came in who got to define such concepts. The Rationalist/Materialist tradition established by Bacon, Newton, and Locke asserted that only information provided by measurement and experimentation could provide understanding of phenomena, which implied that science, as defined by the Western European intellectual tradition, was the only legitimate interpreter of the natural world.

The Enlightenment occurred during the seventeenth and eighteenth centuries, which also happened to be the period during which Europeans encountered Indigenous peoples in many parts of the world, including North and South America and Australia. There was clearly a difference in the ways in which these cultural traditions saw themselves with regard to the natural world. The European (Western) tradition assumed that humans were meant to be dominant over nature. This is ironic, since it arises from the idea of "dominion" in the first book of Genesis, even though one stated purpose of the Enlightenment was to reduce the power and influence of the church over human thought.

In contrast to the Western way of seeing the world, Indigenous peoples tended to view themselves not as dominant over, but as connected to and part of, the natural world. Accepting connection with nature involves less want. Your needs are essentially the same as those of nonhumans, thus it becomes easy to identify with them. Anderson (1996: 86–87) identified such needs for all organisms as (1) food and water, (2) temperature regulation within physiological comfort zones, (3) health and physical safety, (4) sleep and arousal (stimulation), (5) sex and reproduction, (6) social needs, and (7) control over one's own life. Careful consideration of this fairly unremarkable list makes it clear that there is really much less difference than is often realized between man and mouse, or even between humans and insects. It is also apparent that meeting these needs does not require domination of nature.

My family's home life was a lifelong experiment in seeing how these conflicting perspectives might coexist, forcing me to confront these questions at a young age. The way I chose to deal with these questions was to look to the nonhuman world. This was okay for me as child, but by the time I entered high school I figured out that I had made an odd choice for a young boy working out his own identity issues. The few Indian people with whom I interacted seemed to feel science was something that was "not for Indians."

My mother used to become angry because teachers and other people used to say that Indian people were superstitious and their knowledge was basically "fairy tales." Her thinking seemed similar to that described by an Anishinaabe scholar: "It is doubly difficult for men and women who still speak their ancestral languages to disown their own cultural heritage when they can see that it is as substantive as the competing one, and still has meaning, application, and purpose" (Basil Johnston, cited in Russ 1992: viii). She felt that thinking of the animals that lived around our house as "other peoples" was a sophisticated way of viewing the world, and I tended to agree with her.

Advancing through the educational system I began to see more and more of what I felt was unwarranted arrogance on the part of people trained in science. To me many of these people made foolish and even ignorant statements. If questioned such people would often claim that their ideas came from "science." I found that my mother and I were not alone in our impressions of such behavior. The Standing Rock Lakota scholar and professor Vine Deloria Jr. has written:

> Regardless of what Indians have said regarding their origins, their migrations, their experiences with birds, animals, lands, waters, mountains and other peoples, the scientists have maintained a stranglehold on the definitions of what respectable and reliable human experiences are. The Indian explanation is always cast aside as superstition, precluding Indians from having an appreciable status as human beings, and reducing them in the eyes of educated people to a prehuman level of ignorance. Indians must simply take whatever status they have been granted by scientists at that point at which they become acceptable to science. . . .
>
> The stereotypical image of American Indians as childlike superstitious creatures still remains in the popular American mind—a subhuman species that really has no feelings, values, or inherent worth. This attitude permeates American society because Americans have been taught that scientists are always right, that they have no personal biases, and that they do not lie, three fictions that are impossible to defeat. (Deloria 1995)

Although this statement may be somewhat exaggerated, it expresses an essential frustration in the attitude of Indigenous Americans toward the scientific traditions imposed upon them by European colonists in the Americas. My goal in this book is to correct the injustice identified by Deloria and establish that the knowledge, traditions, and concepts of the Indigenous peoples of North America are solid, empirically based, and worthy of being considered a scientific body of knowledge and theory comparable to that of the European tradition.

The book you hold is divided into eleven chapters that fall into two basic categories. The first five chapters lay out the solid scientific foundations and ideas that underlie what I think of as Indigenous science and Traditional Ecological Knowledge (TEK). I consider this tradition to be as powerful as, but different in approach from, Western science. To paraphrase Russ (1992: xxiii): "I offer what follows not in the hope that it will suddenly and unequivocally explain Native people and their attitudes towards nature . . . It is offered to help make the dominant culture aware of how chronically ignorant we are of the very complex world Native people inhabit." Both Native and Western approaches have their strengths, and I suggest how they might be used to complement each other.

In the first chapter I define terms and present the background and concepts that underlie Indigenous knowledge. The second chapter presents how Indigenous people deal with ecology, specifically how they deal with concepts of community and personhood through their understanding of relationships. Chapter 3 further develops the idea of Indigenous ecology by discussing the difference in the way Western and Indigenous cultures conceive of predators and predation and how this shapes cultural traditions. Chapter 4 is intended to be the most technical chapter, but it demonstrates how Indigenous understanding of nature can be not only different, but also more sophisticated and insightful than related ideas from the Western tradition. This chapter introduces the idea that Indigenous traditions actually incorporate evolutionary concepts; an idea further developed in Chapter 5, which addresses the controversy over evolution in the United States in relation to Indigenous concepts of creation and relatedness. In the sixth chapter I describe the history of ecology and evolutionary thought and how it converges on the principles I established in the previous five chapters. As part of this, I incorporate my own experiences and frustrations during my career as the only faculty member in the United States with a joint appointment in ecology and evolutionary biology and Indigenous studies.

The remaining five chapters deal with applications of the themes I have developed in the first six chapters. In Chapter 7 I discuss how nature and ecology are important themes in fiction written by Native American writers, and why this is important in creating a genuine Indigenous form of storytelling that establishes an Indigenous philosophical outlook. Chapters 8 and 9 deal with a couple of controversial issues that link the natural sciences to Indigenous politics and social issues. In Chapter 8 I evaluate and review *The Ecological Indian* by Shepard Krech (1999a), along with other books that purport to evaluate whether or not Indigenous Americans were "ecological" in their thinking. Chapter 9 discusses the "scientific" writings and ideas of the prominent Lakota scholar Vine Deloria Jr., particularly in relation to debates over "Pleistocene (aboriginal) overkill." I have complex feelings about the writings of Deloria. On the one hand I feel that he was one of the most important scholars that Indigenous America has ever produced. Some of his ideas are powerful and continue to be important in establishing Indigenous thought and philosophy in scholarly circles. At the same time I think Deloria never really understood the actual nature of the scientific endeavor and how Indigenous thought could actually be supported by Indigenizing science. As a result, his attempts to critically evaluate science involve numerous errors and alternatives that are not seriously defensible, either in the scientific or the Indigenous traditions.

In Chapter 10 I discuss why academic programs that focus on Indigenous peoples actively avoid including scholars who work in the natural sciences, and why this may be a problem for understanding the contributions that Indigenous people can make to academia in general. Finally, in Chapter 11 I evaluate the overall significance of TEK and explain why I think it presents a valid alternative as well as possible solutions to some of the most pressing problems in contemporary society.

A NOTE ON TERMINOLOGY

When I refer to our nonhuman relatives there are several conventions to which I try to adhere. First, I typically give the scientific (Linnaean) binomial (e.g., *Canis lupus* for wolf) the first time I mention a species as a biological entity. Birds have formal and internationally recognized names, so when I refer to birds by their species name (e.g., Red-tailed Hawk, Western Gull), these names should be capitalized, but when I refer to birds in general (e.g., crows, gulls, puffins), these names are not capitalized. Mammals do not have standardized formal names, so their names are typically not capitalized. The only other circumstance in which a name of an animal is capitalized is when I refer to a character from the traditional stories of Indigenous people or creator figures, such as Wolf, Coyote, or Raven. I apologize if this causes confusion, but hope the criteria I have outlined provide guidance.

There are two other important points that I wish to address. First is the issue of what is described as "pan-Indianism," or the tendency to generalize about the knowledge and traditions of Indigenous Americans that is alleged to blur important distinctions between the knowledge and stories of different peoples or tribal traditions. This supposed critique seems to be employed as a means of deflecting or distracting Indigenous people from arguing that perspectives and philosophical themes exist that are shared by almost all Indigenous peoples, and that these themes yield powerful insights into the functioning of the natural world that rival or even exceed those of the European philosophical and intellectual traditions in sophistication and usefulness. I have never heard it argued that there is a problem with a pan-European body of knowledge. In fact, just the opposite happens—the universality of ideas in the Western tradition is considered to be a strength of this tradition. Charges of pan-Indianism may be employed to prevent such views from being recognized as a powerful and valid alternative to the Western tradition.

I do not contend that different peoples and cultural traditions did not have stories and traditions that differed from one another. North American Indigenous worldviews obviously vary in the details of their belief systems, but much less so in the principles that underlie their philosophical beliefs (Cordova 2007). As the Canadian Salish/Metis writer Lee Maracle states, "We come from our own specific places, but we have a commonality and a common dream" (in Eigenbrod 2005: 83). The influence of local places upon cultures and the corresponding diversity of peoples attached to those places guarantee the existence of variation in the ceremonial and symbolic expressions of Indigenous worldviews. Despite this spatial variation in ecology and physical space there appear to exist a fundamental shared way of thinking and a concept of community common to Indigenous peoples of North America, which I link to TEK. This fundamental shared body of ideas and concepts constitutes a scientific approach that, as with the European tradition, helps establish the overall validity of this approach.

Despite both forced and voluntary relocations, Indigenous people have taken their TEK with them, which has allowed them to survive these experiences and establish sacred places in their new homes. For example, when the Cherokee were forcibly relocated from the mountains of North Carolina and Tennessee to eastern Oklahoma, they developed stories and traditions that were founded on the same fundamental principles but that were specific to the new physical locality (Owens 1998: 164). This way of thought includes (1) respect for nonhuman entities as individuals and the existence of bonds between humans and nonhumans, including incorporation of nonhumans into ethical codes of behavior, (2) the importance of local places, and (3) the recognition of humans as part of the ecological system, rather than as separate from and defining that system. Despite dislocations and forced removals, these ideas are part of the shared intellectual property of all Indigenous peoples from several dozen tribes with which I have worked during the last fifteen years at Haskell Indian Nations University in Lawrence, Kansas, and Northwest Indian College outside of Bellingham, Washington.

My next point concerns terminology used to describe the Indigenous peoples of the Americas. A number of terms have been used over the years, including *Indian, Native, Native American, Aboriginal,* and *Indigenous. Native American* seems to be a term invented in academe in an effort at political correctness (Harjo and Bird 1997). This term was invented in order to replace the term *American Indian*, which replaced *Indian (or Indio)*, which was mistakenly applied by representatives of Spain, including Cristobal Colon, who mistakenly thought they had reached the East Indies, and named the people they found there according to this mistake (Sale 1991).

In Indigenous communities people typically identify themselves first by their tribal name, although even these names are often not the true names they call themselves. In most cases names assigned to tribes by Europeans are based on names given by other tribes, such as Sioux for Lakota and Dakota, Ojibway or Chippewa for Anishinaabe, Cheyenne for Tsistsista, Comanche and Shoshone for Numunu and Newa-nuh (Nuhmuhnuh), Creek for Muskogee, and so on. Ironically these incorrect names have become the official designations for these peoples according to the U.S. government. People are registered with tribes that have names that were rarely, if ever, used by the peoples themselves.

The standard generic terms used to describe Indigenous people are *Indian* in the United States and *Native* or *Aboriginal* in Canada. Canadian tribes also use the term *First Nations* to describe their peoples as a collective group (Harjo and Bird 1997). The term I employ is *Indigenous.* My reason for making this choice is that I am by training a scientist. *Indigenous* means "belonging to a place, originating in and typical of a region or country," and is derived from the Latin word for "born in" (Webster's 1988). As the Ojibway writer Richard Wagemese says, "traditional Indian teachings say that all things, including man, come from the land . . . The Indian, born of the land, had no need to own or control it, because there is no ownership of something of which you are an active part" (Wagamese 1996: 36). To me, this term best exemplifies the concept that these are the people "born of" this land now referred to as North America, although for sake of clarity at times I use the terms *Indian, Native American*, *Native*, and *Indigenous* interchangeably, especially when citing other authors who use these terms.

1 Defining Traditional Ecological Knowledge

> In the old days we learned everything at once, then had to take it apart to understand it. When I went to white school, I had to learn everything in little parts, then try to put it together again. I thought that was backwards. I still do.
>
> Lakota Elder, quoted in Kent Nerburn (2009)

When European immigrants first encountered the Indigenous peoples of the Americas during the sixteenth through nineteenth centuries, the latter almost certainly had a more sophisticated understanding of the basic principles of ecology and possibly even of evolutionary biology (J. Marshall 1995, 2001; see below). The ignorance of the average American of European heritage was quite striking. In 1907 the American philosopher Henry Adams said in reference to the year of his birth, "In essentials like religion, ethics, philosophy, history, literature, art, and the concepts of science the American of 1854 stood closer to the year 1 than to the year 1900" (Adams 1918).

Given the above, it is important to consider the state of knowledge of Indigenous Americans prior to the arrival of Europeans, especially with regard to biological phenomena. This knowledge and its relationship to Western or European knowledge, both contemporary and historical, are the topic of this book. Considerable discussion exists in the literature about the nature of biological knowledge held by Indigenous peoples, particularly the Indigenous peoples of North America (Huntington, 1992; Anderson 1996; Hughes 1996; Berkes 1999; Krech 1999a; Pierotti and Wildcat 1997a, 1997b, 1999b, 2000). There appears to be a general consensus that Indigenous knowledge can be important, especially in applied situations, such as agricultural practices and fire ecology (Johannes 1989; Berkes 1999; Mt. Pleasant 2001). Few studies, however, treat the knowledge of Indigenous peoples as equivalent in value to that of the Western scientific tradition, although this is occasionally recognized: "Being keen and vigilant observers, scientists in the broadest sense of the word, Indigenous peoples have not only used the resources around them, but have maintained and enhanced them" (Turner 2005: 14).

Most studies of Indigenous knowledge emphasize the local knowledge bases of individual peoples (e.g., Tanner 1979; Johannes 1989; Kawagley 1995; Anderson 1996; Berkes 1999; Turner 2005; Menzies 2006; Henriksen 2009). In this book my intention is to establish the principles that underlie Indigenous knowledge and make it comparable to fields in Western science such as ecology, where basic principles such as competition and trophic levels underpin field studies that are themselves very local in their data and possible applications. In this effort I hope to establish Indigenous approaches as equivalent to Western science in usefulness and insight, but different in approach.

This last point is crucial because it is important to understand that the knowledge of Indigenous peoples emerges from a different philosophical tradition than what is described in the Western tradition as scientific knowledge. To understand this it is essential to consider differences in worldview. Yupiaq scholar Oscar Kawagley suggests that the way to understand the concept of worldview is to consider the answers to the following questions: (1) What is real? (metaphysics); (2) What can we understand? (epistemology); (3) How should we behave? (ethics); (4) What is pleasing to the senses? (aesthetics); and (5) What are the patterns upon which we can rely? (logic) (Kawagley 1995). Indigenous traditions would answer each of those questions differently than would the European tradition.

For example, in the realm of metaphysics, Indigenous people recognize the significance of unusual or unique events, which the European tradition would characterize as anecdotes rather than data (Mitchell et al. 1997; Bekoff 2000). In the realm of "scientific" knowledge, this means that the European tradition uses the metaphor of a statistically "average" event to carry more weight than the single "unusual" event (see Chapter 4 for a deeper exploration of this topic). Unusual events are often attributed to the activities of "spirits" in Indigenous traditions, although this English term is inadequate to translate the actual meaning of the concept (Marshall 2005). In any case, to Indigenous knowledge traditions such "spiritual" causes or happenings are not seen as "supernatural." This implies that they are not outside the realm of what is real, but part of the natural order of things and thus readily subject to interpretation and understanding (e.g., Brody 1982; Henriksen 2009).

This acceptance of unusual events as part of reality obviously changes the epistemology of a culture. Thus, what scientists of European heritage consider a "demon-haunted world" (Sagan 1995; Pavlik 1997) was, in fact, a world full of influences, some of which were well understood, others of which were less well understood, but all of which were comprehensible and could be dealt with (Henriksen 2009).

With regard to ethics, Western thought has traditionally followed the lead of Aristotle and defined politics and ethics as occurring exclusively within the human realm. Aristotle proposed that human values are learned from our fellow community members. From the Indigenous perspective Aristotle's basic reasoning was right, but his notion of community membership was wrong in that it was insufficiently inclusive, excluding the non-human portion of the community (see Chapter 2 for further exploration of this theme). Indigenous thought defines politics and ethics as existing in the realm of ecological communities and ecosystems. Following this logic it makes no sense to limit the notion of politics and ethics only to human beings. By limiting the definition of "persons" to human beings, however, Aristotle created an overly narrow sense of community and the corresponding spheres of political and moral life.

Indigenous knowledge emerges from careful long-term observation of natural phenomena. The "data" collected using this approach are basically an understanding of relationships between specific biological entities (plants, mammals, birds, insects, fish) and among biological and physical entities (rocks, bodies of water) landforms (mountains, isolated hills, unusual rock formations), and meteorological phenomena (Barsh 2000). This knowledge encompasses practical, empirical, and ideological aspects of understanding and is both the information itself and a way of knowing (Mailhot 1994; Howitt 2001; Henriksen 2009).

These observational "data," or information obtained through such approaches, are so precise and thorough that they are constrained to focus on local environments or "places" (Swann 1991; Kidwell and Velie 2005). One definition of Traditional Ecological Knowledge is "the sum of the data and ideas acquired by a human group on its environment resulting from the group's use and occupation of a specific region over many generations" (Mailhot 1994). It is only possible to know a limited area in the kind of detail required for true Indigenous knowledge. Thus by definition many of the specific results obtained can only have local application (Brody 1982).

In contrast, the Western "scientific" tradition seeks "global" solutions—that is, results that can be generalized across all localities, or "so that it could be used by groups of people who did not necessarily live in the same region" (Alessa 2009). This can create problems in that solutions and results that are assumed to be global in scope turn out instead to be local. For example, when I was an undergraduate in the early 1970s I listened to an endless, sometimes acrimonious debate between two graduate students, one of whom studied Steller sea lions, *Eumetopias jubatus*, in Alaska, whereas the other studied the same species in California. The investigator who worked in Alaska insisted that parental care lasted for more than a year in this species, whereas the California investigator insisted with equal assurance that offspring were weaned at the age of three to four months. Both insisted that their view of parent-offspring relationships in these sea lions was correct and that the other must be wrong. When I suggested that they both might be right and that ecological conditions in different locations might require different responses, both investigators dismissed me as a naïve undergraduate who "did not understand how science worked."

In the long run it turned out that both views were correct. Harsher conditions in Alaska favored extended parental care, whereas milder conditions in California allowed sea lions to wean their young at younger ages. To me as a larval-stage scientist, this debate revealed the limitations of the Western typological, single "global solution" approach to science. The irony, of course, was that given the proclivity of Indigenous peoples to accept unusual observations and incorporate them in their understanding of the world, if an Aleut from Alaska had met with a Yurok from California and presented these different results, both would have completely accepted the statements made by the other as factual, and both would have presented solid explanations based upon their knowledge of local environmental conditions for why this should be the case.

The important point to take from this example is that an individual's worldview—the way that this individual sees the world—has a major impact on the way that he or she interprets it. To further illustrate this point, it is apparent from studies by Japanese and Western investigators that the basic perspectives of these two cultural traditions influence both the types of questions asked and the interpretations of observations collected (Asquith 1997). To the Japanese, emotionality rather than rationality is central to their way of distinguishing between human and nonhuman (Ohnuki-Tierney 1995). In contrast, the Western tradition believes in "objective," "rational" distinctions and feels that the personalities of individual animals cannot be quantified. Thus, questions into such topics remain outside the boundaries of proper scientific inquiry (Asquith 1986, 1996, 1997; see also Chapter 4). This attitude on the part of Western science resulted in a lag of more than two decades behind the Japanese in key theoretical developments in the study of primate behavior (Asquith 1996, 1997).

In the intellectual and philosophical traditions of the Indigenous peoples of North America no question is off limits, but there are some things that probably "cannot be known." The observations on which knowledge of a local system is based are so careful and detailed that any aspect of relationships can be discussed, among humans, between human and nonhuman, and among nonhumans (see also Barsh 2000). It is crucial to keep in mind, however, that this knowledge would be applicable only to local humans and nonhumans—that is, those who came from and shared the same "place" (Swann 1991; Anderson 1996; Basso 1996; Kidwell and Velie 2005). The knowledge held by Indigenous people is specific, which means it is also very accurate and may often be superior to Western science in its ability to predict local phenomena. Both are valid forms of "knowing," but the scale at which they can be applied differs (Alessa 2009).

Despite this emphasis on "local" knowledge, there are similar philosophical and conceptual themes within the knowledge of different groups of Indigenous peoples (Pierotti and Wildcat 2000). These similarities can be seen in the "ideological aspects" (Mailhot 1994) or "ways of knowing" (Howitt 2001) that have been described. Shared themes emerge from a shared understanding of the nature of relationships and relatedness that exist within and between social and ecological communities (see Chapters 2–4).

Along with debates concerning the possible significance of knowledge held by Indigenous peoples, there has also been discussion about which of a variety of terms is most appropriate to characterize such knowledge.

Among the candidates that have been advanced are *Traditional Ecological Knowledge* (TEK) (Johannes 1989; Berkes 1999; Pierotti and Wildcat 2000), *Indigenous Knowledge* (Warren et al. 1995), *Local Knowledge* (Berkes 1999; A. Tanner, 18 April 2005, personal communication), *Folk Knowledge* (Alessa 2009), and *Naturalized Knowledge Systems* (Henry Lickers, personal communication).

Each of these terms approaches the issue of how to conceive of such knowledge, but none seems to satisfy everyone. Of these terms, Indigenous Knowledge seems to be most inclusive, but it may be overly broad, especially because its use in this context appears to assume that all knowledge of Indigenous peoples is concerned with natural phenomena (Berkes 1999). While it can be argued that this is true in a vague general sense, it is also misleading because Indigenous people have a wide range of knowledge that does not deal specifically with the natural world.

Local Knowledge underscores the fact that virtually all such knowledge is strongly tied to specific locations where the people under consideration make their home and is specific to their way of life (Deloria 1992; Basso 1996; Pierotti and Wildcat 2000). As such, Local Knowledge is a useful term, but fails to acknowledge the general themes and attitudes that generate a "way of knowing" shared among a wide range of peoples with regard to knowledge of the natural world. In addition some have argued that the concept of Local Knowledge can also be applied to non-Indigenous communities. Similar problems can be found in the concept of Naturalized Knowledge Systems, which has been deliberately designed as a concept that may be applied to any cultural tradition (H. Lickers, personal communication).

Traditional Ecological Knowledge (TEK) is my term of choice and will be the primary term employed in this text. It has been argued that use of the word *traditional* implies that such concepts or knowledge are of the past and unchangeable, and therefore may not be relevant to the contemporary world. In fact, it has been stated by non-Indigenous investigators that *traditional* and *change* are contradictory concepts (Berkes 1999). In contrast, I believe that although use of the term *traditional* clearly implies that such knowledge and its related concepts have been in existence for a considerable length of time, the reason for such longevity is precisely because their ability to incorporate new observations and information has kept them fresh and relevant.

Problems arise with the concept of *traditional* primarily because the Western philosophical and academic approach assumes that all established traditions must be written down and hence unchangeable, such as the Bible or Darwin's *Origin of Species* (Darwin 1859). It has been argued by some students of this topic that "the term [*traditional*] also carries the generally unacknowledged connotation that the item in question is now in decline, and thus is in need of being 'preserved,' such as by placing it in a museum" (A. Tanner, personal communication, 18 April 2005).

Tanner also says that "[use of the term *traditional*] suggests that our interest in aboriginal knowledge is essentially limited to those aspects that have existed within a group for many generations. It thus implies a focus on ideas that are *'ghettoized' in the past*. The second reason to avoid the term 'traditional' in the phrase TEK is the implication of the English term, which suggests origins in the dim and distant past, and which, for this reason, involves knowledge that is not open to *normal processes of verification*" (A. Tanner, personal communication, 18 April 2005; emphasis added).

To me, the choice of words here is very instructive. I do not think that TEK involves knowledge that "is not open to *normal processes* of verification." As I suggest above, at the local level TEK may be better at predicting outcomes than the more general models and metaphors of Western science.

In non-Western cultures, especially those based primarily on oral rather than written traditions, new information and ways of thinking are regularly incorporated into traditional concepts. The problem Tanner poses may not really be a problem, because such knowledge is continually subject to verification. Information that is no longer valid or information that can be replaced by newly acquired information is no longer considered to be important by Indigenous peoples. Tanner recognizes that this takes place: "in my experience, aboriginal people are continually checking on the validity of knowledge they share with others, *in the same way scientists do*" (emphasis added).

A more perplexing perspective has been raised in which it is said that, "(traditional knowledge) is not really 'knowledge' at all; it's more a way of life" (Kluane First Nation member quoted in Nadasdy 2003: 63). I understand the point being made, but I think that this is an example of talking past one another. The issue that Nadasdy (and his Kluane colleagues) seem to want emphasized is that to the Kluane people, hunting is a way of life, and that hunting consists of everything from the first thoughts about when to start, through the kill and the ultimate preparation, on to ultimate allotment of the "meat" that is gathered as a part of the hunting process. I think that two different forms of knowledge are being discussed here: 1) patterns that have been shown to work over many generations which are the equivalents of premises or postulates in Western science (e.g., trophic dynamics), and 2) specific observations that are incorporated into these patterns (e.g., notation that one species preys upon or is preyed upon by another at certain times of the year). This can be illustrated by an example used by Nadasdy when discussing an elder and superior hunter named Moose Jackson. "Moose said he can tell by looking at moose tracks how old they are . . . where the sun was at the time the moose was there and the direction the wind was blowing. This information, *combined with his knowledge of moose behavior*, is enough for him to know where the moose was going and what it was doing . . . (and) is enough to tell him where the moose is . . . he does not actually have to follow the moose tracks, he merely goes to where he knows the moose will be" (Nadasdy 2003: 107, emphasis added).

This is very detailed and local knowledge, but it is based upon careful empirical observation. It is also crucial that Jackson has previous knowledge of moose behavior. I can readily imagine an experienced field biologist from the Western tradition operating at a similar level of sophistication (see Chapter 6). The main difference would be that the biologist would write down information about wind direction, depth and symmetry of tracks, sun angle, etc. whereas Moose Jackson has this information carefully stored away. There would also be a difference in the way this material is passed on. The biologist would publish the most substantial of their findings, give talks at meetings, and perhaps train student assistants, but they would not include every detail of the activities that they employ in the field. In contrast, Moose Jackson would take younger hunters out and show them directly, and would tell them stories of previous moose hunts. The two approaches are quite similar in practice, but the biologist is paid a salary and buys his "meat." Therefore, this is not seen as a way of life in the sense that "the single most important aspect of Kluane people's way of life is hunting" (Nadasdy 2003, p. 63).

What is ignored in many discussions of traditional knowledge, ecological or otherwise, is that the stories, the oral tradition, are only the foundation of the actual knowledge base, and functionally equivalent to basic concepts or premises, such as MacArthur's "Broken stick model" in interspecific competition (MacArthur 1972) or life tables in population ecology. Younger people are told the stories, which provide a framework for the basic knowledge set to which they can add new insights as they gain experience. This is comparable to the way effective educators of the dominant culture do not force their students to learn every fact, but instead provide a framework within which new information acquired by each individual can be placed (see also Kawagley 1995; Preece 1999). Thus, Indigenous stories function as metaphors that allow a philosophical framework to develop (see Chapter 4). It is not that stories are considered to be literally true, any more than models of exponential and logistic population growth are expected to literally represent the actual behavior of animal populations. Like mathematical models of population phenomena, traditional stories provide a context into which empirical observations can be placed and compared against the assumed state. TEK can be rebuilt and changed by each generation in the same way that each new generation of graduates in ecology or evolutionary biology go out and make new discoveries that add on to the base of empirical knowledge that currently exists in those fields.

The tendency of Indigenous societies to incorporate new information, and new technologies, is crucial. Unfortunately, adding new ideas, and especially new technologies, has led to debate over whether or not use of modern technology, such as guns or outboard motors, removes hunting and fishing practices from the category of "traditional" in terms of treaty arrangements. Indigenous people invariably contend that the attitude and philosophy involved, rather than the technology, are what make a practice traditional. In contrast, many of European heritage argue that traditional practices can employ only primitive technologies, such as bows and arrows, presumably using stone arrowheads, or canoes, presumably dugout or manufactured from birchbark.

Following such logic to an absurd extreme, it might be questioned whether the buffalo cultures of the Plains peoples of North America, including the Comanche, Cheyenne, and Sioux peoples, can be considered as traditional. Hunting from horseback was a major part of this tradition, yet it is well established that Europeans introduced domestic horses into North America in the sixteenth century (Isenberg 2000). Prior to this the only domestic animal of plains peoples was the wolf (or dog, *Canis lupus*). These dogs were almost pure wolf as described in both Euro-American accounts (Hyde 1968: 9–11; Morey 1986) and in traditional stories by Indigenous elders (H. Mann, personal communication).

It seems likely that most people of European ancestry assume that *traditional* describes only those conditions that existed when a tribe was initially visited and first described by Europeans. This contains the tacit assumption that Indigenous peoples remained essentially unchanged and uninfluenced by any other cultures prior to European contact, and is the social equivalent of a creationist perspective (see Chapter 5). This is, of course, absurd—tribes regularly exchanged materials and ideas with one another for millennia prior to 1492 (C. Martin 1978; C. Mann 2005). For example, southwestern tribes regularly incorporated shells of marine mollusks in their decoration and in barter, yet the nearest source of shells was either the Pacific Coast or the Sea of Cortez, both several hundred miles away across country that was primarily hot desert.

Use of the term *ecological* to describe such knowledge is important because it establishes the links of these flexible and adaptable traditions to the natural world. At the same time, TEK is not so inclusive that it appears to incorporate all of the knowledge held by Indigenous peoples. Acknowledging the ecological component of such knowledge, however, establishes that TEK is scientific in most meaningful senses of the word. That is, TEK is based on empirical knowledge that has been collected over long periods of time and incorporated into an organized way of understanding how the world functions based on relationships observed and understood at a local scale.

SEPARATION OF KNOWLEDGE FROM SPIRIT IN THE WESTERN TRADITION

Within the Western knowledge system there is a "separation of areas called science from those called art and religion. The Indigenous knowledge base on the other hand integrates those areas of knowledge so that science is both religious and aesthetic" (Cordero 1995). This suggests that although

TEK is scientific, it differs philosophically from Western science. One point consistently missed by Western investigators of TEK is the significance of honor and responsibility in Indigenous traditions, which lend TEK its "religious" or "spiritual" overtones. Some students of TEK consider this a problem, whereas, as argued above, I consider it a strength. In the Western tradition, it is considered to be a problem that TEK combines empirically based observations and conclusions with spiritual/religious/philosophical ideas. It is assumed that to set "Western scientific knowledge" against TEK, it is necessary to specify and acknowledge an equivalent distinction within Indigenous knowledge between those aspects that are more or less equivalent to "science" and "religion."

This way of thinking contains several hidden assumptions on the part of many adherents of "Western science." To begin with Western "science," as it currently defines itself, is a relatively recent philosophical development (Alessa 2009). In the early part of the nineteenth century the approach to the study of nature was defined by Romantics such as Thoreau and Goethe, which assumed a correspondence between the inner nature of humans and the structure of external reality, that is, man reflects nature's order (Worster 1994: 89). By the middle of the century, however, the scientific community decided to pursue (or at least attempt) a different epistemological direction, setting an ideal of detachment and objectivity. Science concerns nature, but it is also about things, which can only be analyzed, measured, and numbered. Thus, these scientists attempted to exclude any approaches that *bore the taint of personal vision* (Worster 1994: 90, emphasis added), combined with refusal to accept the Romantic notion of correspondence and its ethic of knowledge, because they were unsure what Romantics meant by inner spirit and how that spirit might be related to matter. Nonetheless, Western science is still influenced by the culture from which it emerges because "no body of knowledge exists for its own sake outside the moral framework of understanding" (Cajete 2000: 76).

There has been a tendency in some circles to equate Indigenous philosophical traditions with the Romantic Movement because of the idea that the human spirit reflects nature's order. This assumption is naïve at a couple of levels. First, as I shall argue throughout this book Indigenous people do not assume a simple order in nature. Second, and more importantly, the Romantics did not really regard themselves as part of nature. As an example, Thoreau wrote, "Man cannot afford . . . to look at Nature directly, but only with the side of his eye. He must look through and beyond her" (McIntosh 1974; Worster 1994: 98). Thoreau thought it was necessary to transcend the natural world and strive for a supernatural life, a tendency that increased as he followed Emersonian transcendentalism, which argued that spiritual needs cannot be met by the material world (Worster 1994). Thoreau and Emerson tended to look beyond nature to a grander sphere of ideal forms, which seemed to draw directly on the concept of Platonic "ideals," a tendency that can also be seen in the writings of other Romantics, such as Blake and Coleridge. In the long run, Thoreau's transcendental idealism directs attention away from the natural order and towards an Eden-like utopia (McIntosh 1974; Worster 1994).

This is quite different than Indigenous epistemology (Kawagley 1995). Thoreau contends that "Man cannot afford . . . to look at Nature directly" (McIntosh 1974), however, this is exactly what Indigenous people do, try to understand nature exactly as how it is so they can figure out how to survive and prosper within its changeable actions. As a result, they transcend transcendentalism and fulfill their spiritual needs through their interactions with the natural world (see Chapters 2–5). Finally, rather than seeking a higher moral law, as did Thoreau and Emerson, in the Indigenous tradition morality and ethics are defined through ceremonies linked to the natural world and their relationships to it (see Chapters 3 and 5) and the wisdom of elders, which is acquired through life long interaction with nature.

PHILOSOPHY AND METAPHOR IN UNDERSTANDING NATURE

If we consider that spiritual equals philosophical as many contemporary scientists might argue, then the Western scientific tradition is full of such ideas and language. To list a few: Cartesian dualism and use of the machine metaphor to model nonhuman organisms, Platonic ideals in typological species identities, and the tendency to emphasize competition over cooperation in understanding ecological relationships. The Western scientific tradition depends heavily on models and metaphors that have Western philosophical or economic ideas at their roots (see also Worster 1993, especially Chapter 16; Mitchell et al. 1997, Lewontin 2001).

As an example, the late Oxford evolutionary biologist W. D. Hamilton has made the statement that, with regard to eugenics, "human directed selection, whether to *maintain standards* or to *speed the intellectual and physical progress* of humanity, could be made both more effective and more merciful than the *obviously inefficient and cruel natural process*" (Hamilton 1996, 15; emphasis added). Hamilton is the author of one of the most widely cited "models" in evolutionary biology, that is, the idea that altruistic or cooperative behavior among both humans and nonhumans is driven by genetic relatedness or "kin selection." Even though Hamilton was not a believer in eugenics, his model is often cited as an example of genetic "determinism" (Hamilton 1996). It is also instructive that this model has proved problematic as more information is gained concerning the social interactions both within and among species (see Chapters 2 and 4).

More to the point, Hamilton's phrasing emphasizes separation of humans from natural processes, a point discussed in greater detail later in this book. In a comment directed at both Hamilton and his intellectual descendant Richard Dawkins, the evolutionary biologist Gabriel Dover has stated with regard to models, "the use of coefficients [of selection] . . . does not mean that this is an adequate representation of what is happening in nature; it is *merely a simplifying device to make the mathematics more tractable*. Organisms evolve under very complex internal and external conditions which have not so far been captured by available models" (Dover 2000; emphasis added).

In its alleged impersonal "objectivity," Western science typically fails to consider the responsibility scientists owe toward study organisms. There is also little, if any, discussion of the obvious reciprocal relationship between investigator and subject. As described by a major philosopher from the Western scientific tradition: "Whatever the scientists' feelings, or style, while working, these are purged from the final work" (Wolpert 1992: 57).

In contrast, Indigenous traditions consider such reciprocal relationships as central to understanding the basis by which knowledge has been acquired through careful and detailed observation (see Trosper 2009 for an extensive discussion of this topic). The Osage scholar John Joseph Matthews has stated that "Osage religious concepts are intertwined with nature through three principles of life: 'self preservation, the necessity of reproduction, and a Force that inspires a bird to sing for the sheer joy of singing'" (Logsdon 1972: 74). Indigenous belief systems have also been described as: "The Native world should be understood as one of multiple communities of sentient beings in a variety of corporeal forms . . . the (important question is) what did experience teach them about reality and how they adapted their lives and beliefs to these lessons?" (Dreyfus 2008: 20–21). In such thinking, indigenous religious traditions seem very close to Darwinian principles, especially in that such traditions accept that humans are not very different from other species.

Spiritual/religious aspects of TEK emerge from attempts to comprehend the nature of a variable and somewhat unpredictable environment and efforts to establish covenants with the natural world that were designed to reduce the negative impact of human actions. These are a form of science, sort of the equivalent of "resource management" but among "relatives," rather than between "exploiters" and "resources." The ceremonies and rituals are linked to knowledge and science in the same way that research protocols are designed to minimize negative impacts on study populations. If we consider religion as described by Durkheim (1961), that is, the way in which a society codes its ethics through ritual, then Indigenous spiritual actions incorporate ethics without invoking threats of divine punishment. Indigenous belief contends that if you mistreat an entity, that entity itself may respond in a way that can cause harm to you and your people, which implies a direct interaction between humans and some other entity similar to the dynamics between predators and prey. In this they differ from the Romantic Tradition, where it was assumed that humans cannot look at Nature directly, but through Nature in order to find a deeper meaning (above).

It is important to emphasize that Indigenous Americans are not that different than other human beings, except that they seem to take the concept of responsibility as a cultural norm more seriously than do most cultures. Because they look at Nature directly, they understand relationships in a more personal fashion than is generally accepted in contemporary Western thought and understand the functioning of natural systems in terms of relationships (Anderson 1996; Barsh 2000). Indigenous Americans extend the concept of community to whom responsibility is owed to include the nonhuman (see Chapters 2 and 3). Honor and responsibility permeate all aspects of Indigenous life, including interactions with the natural and nonhuman world.

At this point, it must be mentioned that many contemporary Indigenous individuals have either not been taught such traditions or may have even rejected them in favor of Western economic approaches. This is a legacy of boarding schools and forced assimilation, a theme dealt with in greater detail near the end of my second and third chapters. Despite the impact of colonization, there still remain two basic premises of Indigenous philosophy, tantamount to religious principles in Indigenous tradition: (1) All things are connected, and (2) All things are related. These premises, discussed in much greater detail in Chapters 2 and 5, respectively, acknowledge the dependence of human beings on all of the other beings with which they share their local ecosystem (Pierotti and Wildcat 1997a, 1997b, 1999b, 2000). These principles share many traits with basic premises of ecology and evolutionary biology and can be found in various forms across a range of Indigenous cultures (Pierotti and Wildcat 1997a, 1997b).

An essential difference between Western and Indigenous perspectives can be illustrated by the Onondaga elder Oren Lyons, who stated, when discussing perceptions of the natural world: "Every time we talk about the nonhuman world we always talk about relatives, whereas you [individuals of European ancestry] always talk about resources" (D. Wildcat, personal communication). Thus, relatedness is seen to be more important than the economic aspects of the relationship between human and nonhuman.

Another way of illustrating the importance of relatedness and connectedness between human and nonhuman can be found in the stories that appear in most Indigenous cultures about how in earlier times animals were considered equivalent to people and both "spoke the same language." Such stories are invariably construed by investigators of European heritage as anthropomorphic constructions in which animals behaved and dressed like humans. A classic example of this can be seen in Barry Lopez's charming tale *Crow and Weasel* (1990), which portrays animals dressed in clothing of Indigenous peoples, speaking human languages, and showing human cultural attributes.

Ignored in such portrayals and interpretations is an equally valid but virtually unacknowledged alternative based on precepts I raised earlier—that such stories are simply a framework within which more detailed empirically derived knowledge about relationships can be placed and evaluated. It can be argued that in ancient times humans probably had much closer relationships with nonhuman animals and were thus much more aware of similarities between human and nonhuman. As a consequence humans understood the ways in which animals communicated with one another both within and among species, including humans. Such ideas are implicit in the arguments presented by Nadasdy (2003), especially in elders, such as Moose Johnson (above). It was not that animals acted and dressed like humans, but instead that humans paid such close attention to their nonhuman relatives that they understood the information being exchanged or transmitted.

This possible difference in interpretation illustrates an important perceptual problem with Western perspectives I discuss in Chapter 4: that communication between humans and non-domestic animals in the Western scientific tradition invariably assumes that animals will learn a human language. This particular, and odd, form of bias can be seen in all "scientific studies" that investigate the ability of animals to communicate with humans, in which animals are taught various human forms of communication, including sign language, particularly in great apes such as chimpanzees, *Pan troglodytes* (Fouts 1973), or gorillas, *Gorilla gorilla* (Patterson 1980; Patterson and Linden 1981); human speech as in African Grey Parrots, *Psittacus erithacus* (Pepperburg 1981, 1983); or commands to nonhumans being given in human languages such as with bottlenose dolphins, *Tursiops truncatus* (Herman 1986). It would seem obvious that the true indicator of higher status would be the ability to understand and communicate in the language of the other. The possibility should be considered that Indigenous peoples learned to communicate with nonhumans by understanding their forms of communication—for example, following birds such as Ravens (see Chapters 3 and 4) or Honeyguides (*Indicator indicator*) to sources of food.

Ironically, many Indigenous peoples in the contemporary world are unwilling to acknowledge this possible interpretation. Another consequence of assimilation and the denigration of Indigenous knowledge traditions, as discussed in the preface by Deloria, is that some young Indigenous people feel, with some justification, that acknowledgment of close affinity of Indigenous people with nonhumans could be used to identify them as being less than human as well (J. Potts, personal communication). This is, of course, a trap, built into the idea that the Western perspective emphasizing this separation from, rather than connection to, the natural world is the watermark indicative of "advanced" or "superior" cultures and is in some circles used as the definition of *civilized* (Coates 1998).

USE OF STORY AS METAPHOR TO UNDERSTAND RELATIONSHIPS

A similar theme can be seen in the use of animal metaphors to designate what are considered to be lesser or inferior beings. One particularly noxious example of this is the way the term *predator*, which is derived from the Latin word for *plunder*, has been used to refer to carnivorous animals, who are not thieving, cruel, or destructive, but simply fulfill a crucial ecological function (see Chapter 3). In a similar vein, humans showing cruel or viciously destructive behavior are described as acting like animals or beasts, whereas such behavior is extremely rare in the nonhuman world. Native American traditions recognized that predatory animals were not cruel; instead they were considered as the beings most similar to humans, who were also recognized as predators (Pierotti and Wildcat 1997b, 2000; Chapter 3). Such recognition was codified through clan names, such as Bear, Wolf, Orca, Eagle, in many tribal cultures that were used to symbolize ecological relatedness (Pierotti and Wildcat 1997a, 2000; Chapters 3 and 4).

In Indigenous traditions it was often considered to be a positive attribute to behave like nonhumans, for there was much to emulate in the nonhuman world (see Chapters 2 and 3). For example, the familial devotion and tenacity of wolves, the fortitude and courage of bears, the strength and stoic nature of bison, were considered to be virtues that humans were pleased to have (Nerburn 1994; J. Marshall 1995).

One key attribute of the stories of Indigenous Americans is that they function both as information about ecological and evolutionary relationships and as instructions about ethical and moral behavior because they emerge from an understanding of relationships between species (Barsh 2000). One classic example of this dual function is the relationship between wolf and coyote in the tales of Numic peoples, including Shoshones and Comanches. To begin with, wolf and coyote are described as brothers, which establishes the evolutionary relationship and shared ancestry between these two closely related species within the genus *Canis* (Lily Pete, in A. Smith 1993: 3; see also Ramsey 1977, Papanikolas 1995). Second, both wolf and coyote are also characterized as good hunters, although wolf is considered to be a much better hunter. For example, wolf is capable of taking adult deer whereas coyote takes primarily fawns, along with rabbits and rats (Lily Pete, in A. Smith 1993: 5; Johnny Dick, in A. Smith 1993: 91), which accurately characterizes the ecological roles of these two species. Thus the ecological and evolutionary relationships among species are a crucial part of the story.

From an ethical and moral perspective wolf and coyote frequently argue about how the world should function, with wolf desiring an idealized world in which death is only temporary, childbirth is easy and pleasant for women, and winter does not exist. In contrast, coyote thinks death should be permanent, childbirth should be difficult, and hardships and cold weather should be regular aspects of human experience (Lily Pete, in A. Smith 1993: 3–4; see also Ramsey 1977, Papanikolas 1995). An example of this can be seen in this story told by Shoshone elder Corbin Harney:

> You see, the coyote and the wolf were talking long ago. Wolf was arguing that we should all look alike, the rocks should be the same, the sagebrush the same, the humans the same, and all the living things on this planet should be the same. We should think alike and act alike and so forth.
>
> But Coyote always said, "No, we should be all be different. We shouldn't look alike at all."
>
> And so today we look around us and nothing looks alike. Rocks are not alike. Humans are not alike. This is the root of why we don't believe in each other. It's just as Coyote said. There's no use believing in just one thing. Let's not believe it. Let's all disagree, and everybody believes in different things. That's why I always say, it's easy to believe the bad things first, but the good thing is harder to believe and harder to come by. As Wolf said, "It's going to be really hard that way, because what you're saying is, let's not believe in each other."
>
> So today, what Coyote said is what we've got . . . (Harney 1995: 26)

The interesting aspect of this dichotomy is that although children are taught to emulate wolf and view wolf as a much more sympathetic figure than coyote, it is coyote who presents the more realistic view of how the world truly functions. In fact, it is stated about coyote's views of death and hardships that "if it weren't for coyote there would be too many people now" (Lily Pete, in A. Smith 1993: 3), which shows a recognition of the risks of local human overpopulation on potentially limiting sources of food, water, and other resources. On those occasions when coyote gets his way, he almost invariably regrets the consequences of his actions and the situations that he originally argued should be the proper state of things. Coyote is shown to be vain and foolish in that he expects the arguments he makes about how the world functions to apply to everyone other than himself. Thus humans are shown that in their own vain, selfish, and ego-driven behavior they act more like coyote, causing problems for themselves and the rest of the world, even though they are expected to try to emulate the idealism and good behavior shown by wolf. Humans are trapped by the real world and resent the sorrow and problems it presents them, but these conditions are attributed to be the result of situations that humans have created (Ramsey 1977; Papanikolas 1995).

The Western tradition seems to have a more extreme and less nuanced dichotomous view of the natural world in which nature is either sentimentalized or treated as if it were cruel and destructive (Coates 1998). In contrast, TEK incorporates a more subtle and nuanced understanding of the functioning of the natural world. Carnivores are recognized as being powerful creatures, not unlike humans, and in the case of wolves are very similar to humans in the structure of their family units. In contrast to wolf, coyote is more of a loner and thereby driven by only his own concerns rather than those of his family. This self-centered behavior causes problems, which can be somewhat alleviated through humans (or wolves) functioning as an integrated group where individuals work for the good of the group rather than individual ends.

From the Indigenous perspective, animal societies are considered to be composed of individuals who are variable in their behavior, especially as environmental conditions change, not just seasonally, but from year to year and decade to decade. This stands in contrast to the Platonic philosophical perspective, which assumes that all animals of a species are fundamentally identical, and the Cartesian philosophical perspective, which assumes that nonhumans are the functional equivalent of machines. As a consequence, Indigenous TEK is set up to understand the functioning of animal societies and the dynamics of animal and plant populations in a very detailed manner. Only in recent years have Western scientists been able to publish similar ideas in the peer-reviewed literature.

This perspective on the part of Indigenous peoples led to ways of interacting with the natural world that minimized the chances of negative impacts of humans upon their environment (Anderson 1996; Kay and Simmons 2002). This is not to say that Indigenous peoples and their activities had no negative impacts, but these impacts were primarily at the level of individual organisms rather than populations or ecosystems. Indigenous people depended upon the taking of the lives of prey animals such as bison, deer, moose, salmon, elk, and beaver, as well as on taking plants for both food and medicine (e.g., prairie turnips, *Echinacea*) or for housing and construction (e.g., birch and cedar bark or logs for longhouses). What was important was that philosophical and spiritual traditions were developed that minimized excessive negative impact on populations or ecosystems (Ross 1992; Anderson 1996). Thus, like the mathematical models so beloved of Western science, the metaphoric stories of TEK are, to paraphrase Dover (2000, quoted above), "simplifying devices to make understanding of relationships more tractable." Specific complex external conditions may not be captured, but the basic approach is clearly laid out.

SPIRITUAL LINKS AS A COMPONENT OF STORIES AND KNOWLEDGE

I believe that it is important that these traditions and TEK have a "spiritual" component, because this meant that the humans involved had an emotional and ethical stake in those behavioral traditions or rituals that minimized the possibility of significant negative impacts. The Osage anthropologist Francis LaFlesche emphasizes, "The survival [of the Osage people] depended on the continued blessing of Wa-kon-da," the spirit power "which was the source of life" (LaFlesche 1995: 31, 34). To survive, the people needed to make sure that they lived in a manner that minimized harmful impacts upon Wa-kon-da.

The issue of religion is important in discussing TEK in comparison to the Western scientific tradition. Many scientists of European ancestry either believe that religion and science are mutually exclusive and fundamentally oppositional from a philosophical perspective, or at least claim that this is their belief (Gould 1987; Worster 1994). This tradition appears to date to the controversy over "Darwinian" evolution resulting from natural selection, which was (and continues to be) treated as if it were antithetical to the Western religious tradition.

In contrast, Indigenous peoples appear to believe that change, including evolutionary responses to changed environmental conditions, is simply a part of the natural world and linked to spiritual understanding. LaFlesche states, "The Osage saw the cosmos as a highly integrated and unified system in which humans were only one factor . . . also that the cosmos was knowable through observation and reasoning. This is not to say that they thought humans could fully understand the cosmos in all its variety and dimensions. They did believe, however, that humans could and should strive constantly to increase their knowledge about the cosmos. To the Osage knowledge was the key to human survival" (LaFlesche 1995: 30). Virtually identical phraseology could be used to describe the Western scientific tradition. The presence of ceremonies and rituals devoted to placating the natural world lends the ecological understanding of Indigenous peoples a "spiritual" component, a topic explored in greater detail in Chapters 2 through 5.

This holistic perspective has led to the caricaturing of Indigenous peoples of the Americas as "natural ecologists" as if they simply are born with a deep understanding of the natural world (this tradition has been criticized in a highly misleading fashion by Krech 1999a, 1999b; and some authors in Harkin and Lewis 2007; see Chapter 8). Such caricatures ignore the fact that ecological understanding was as dearly earned and difficult to obtain for Indigenous people as it is for any contemporary field ecologist.

The lives of Indigenous Americans depended upon their ability to understand the processes of the natural world and their ability to avoid overexploiting animal and plant populations that were crucial as sources of food and clothing. To these peoples TEK was no more abstract or esoteric than any form of economic or ecological knowledge held by people of European ancestry and its academic traditions (see Ross 1992, for an interesting and insightful version of such relationships).

For Indigenous peoples religious practices and ecological knowledge were personal rather than abstract and through stories they personified reality and discussed it through narrative interaction rather than through computation (Anderson 1996; Barsh 2000). "Myth is . . . an alternative form of science . . . an alternative form of investigation . . . It aims, like science, at perceiving and expressing ultimate truths, but the *hypotheses of myth are framed as stories*, not as equations, technical descriptions or taxonomic rules" (Bringhurst 2008: 64; emphasis added). Their concepts were rooted in knowledge of specific locations and specific animal and plant populations over extended periods of time, including an awareness that environmental conditions fluctuate in unpredictable ways and that perturbed systems do not always return to the conditions that existed prior to the perturbation. Unlike the Romantic tradition as exemplified by Thoreau (above) Indigenous peoples regard themselves as an integral part of the ecosystems in which they live, rather than thinking of themselves as being in control of these systems and as having the ability to restore conditions to their "natural state." Being part of a specific ecosystem created meaning and spirituality derived from their relationships to the nonhuman world. The knowledge held by these peoples and cultures was both traditional and ecological.

One final point that is rarely considered when Indigenous knowledge, cultural traditions, and philosophical traditions are discussed is what these beliefs and ideas would look like had they been allowed to evolve for the past five hundred years without interference from Europeans. It needs to be emphasized that when Europeans and the Indigenous peoples of the Americas first encountered one another in the late fifteenth and early sixteenth centuries, the main advantage held by the former was not superior knowledge but instead consisted of the ability to smelt iron and use explosives (taken from East Asian cultures). Europeans were also given an important advantage by the disease organisms that had made the leap from domestic animals into European populations, and to which the Indigenous Americans had no acquired immunity (C. Martin 1978; Diamond 1997; Pierotti 2004; Chapter 2). These "Guns, Germs and Steel" decimated Indigenous populations and allowed the European way to become dominant in the Americas and other regions of the world (Diamond 1997), but in no way did they imply intellectual superiority. In fact, as argued above, when Europeans first encountered Indigenous Americans the latter were superior in terms of their understanding of the natural world and their political and moral philosophy.

Thus we need to consider what these Indigenous cultures might have become had they been allowed to progress "naturally," without the destructive impact of European disease and religious philosophy. One possible example is the Mayan people of southern Mexico and Central America. The picture of the Mayans emphasized by contemporary anthropologists and environmentalists is that "the Mayans possessed accurate calendars and a knowledge of astronomy; they created brilliant sculpture . . . and architecture almost alien in its symmetry. They reached *the Classic period of their culture* between AD 250 and 850, about the year 900, for reasons no one has ever been able to explain, the Mayans abandoned their magnificent structures and the race *fell into an inexplicable decline*" (Cahill 1993: 169; emphasis added).

In contrast to this model of tragedy and disaster is the following story: "The collapse of Maya civilization around 800 to 900 AD seems to have been due to escalating war . . . scorched earth policies did the real damage. However, exhaustion of soil and forest by ordinary agriculture may have been a factor in the collapse. If so, the Maya seem to have learned something . . . Maya strategies do better than most. Where the Maya are poor, no one else can farm at all. Where others survive, the Maya are successful and prosperous" (Anderson 1996: 79–80). The Maya of today have adjusted and may represent the future. They are a people who had an advanced civilization and survived its collapse by returning to traditional ways and applying their own brand of TEK. Today there are as many Maya as there were during the so-called classic period of their culture. The question we probably need to ask ourselves in the context of this book is, "Which is the true classic period of the Maya?"

2 All Things Are Connected

Communities as Both Ecological and Social Entities in Indigenous American Thought

> The great fault of all ethics hitherto is that they believed themselves to have to deal only with the relations of man to man. In reality the question is what is his attitude to the world and all life that comes within his reach. A man is ethical only when life, as such, is sacred to him, that of plants and animals as well as his fellow man, and when he devotes himself helpfully to all life that is in need of help.
>
> Albert Schweitzer (1933)

Community is a concept with numerous meanings or interpretations. The most commonly used definition in contemporary American society is "a group of humans residing in the same locality and under the same government" (*Webster's* 1988). Communities may also be defined as a group or class having common identity, or as sharing a likeness or interest. This can be considered as the social or political concept.

There are also definitions that deal with what is referred to as an ecological community—"a group of plants and animals living in a particular region under more or less similar conditions" (*Webster's* 1988). A more sophisticated definition of an ecological community provided by an ecologist is, "an association of interacting populations, usually defined by the nature of their interaction or by the place in which they live" (Ricklefs 1998). It is clear that under the Western scientific tradition the concept of an ecological community—an interacting assemblage of species—is kept distinct from the concept of a social or political community, consisting entirely of human beings, that is the preeminent meaning of this term in Euro-American cultural traditions (Deloria 1990; Pierotti and Wildcat 2000).

My goal in this chapter is to introduce (or return) the reader to an earlier definition of community, one in which human beings are considered to be part of ecological communities and nonhumans are considered to be part of social communities. In the cultural traditions of the Indigenous peoples of North and South America the distinction between social and ecological communities is not clearly delineated. In these traditions humans regularly have had social interactions and maintained social relationships with plants, animals, and features of the landscape, including rocks, rivers, and mountains (Allen 1986; Deloria 1990; Klubnikin et al. 2000; Tinker 2004; Dreyfus 2008).

This alternate concept of the meaning of community does not arise because Indigenous peoples fail to recognize human social communities, but is instead the result of, first a different concept of what constitutes "personhood" (Pierotti and Wildcat 2000), combined with a tendency to be place or locality oriented such that nonhuman entities that occur in the same ecological area are considered to be more closely related in a functional sense than are unfamiliar human beings (Deloria 1992; Anderson 1996; Basso 1996; Barsh 2000; Pierotti and Wildcat 2000; Kidwell and Velie 2005).

One important reason for these differences in philosophy is that Indigenous peoples lack an immigrant experience within their memories; they assume that they are truly indigenous, that is, born of this land. Native American stories do not deal with the exact time when "historical" events occurred since many such events happened so long ago that they exist "on the other side of memory" (J. Marshall 1995; Pierotti and Wildcat 2000). "Through at least the past 11,000 years, and some say much longer, the land has supported communities of people who have relied on the plants and animals of their home regions for survival. These people have since time immemorial for them, adapted their lifestyles to the changing climates and the fluctuations in abundance of fish, wildlife, and plants" (Turner 2005: 13–14). Although Turner refers specifically to the Indigenous peoples of British Columbia, these principles can be applied to other Indigenous groups as well. In fact, I could cite virtually any detailed description of the belief systems of Indigenous Americans and come up with similar principles. The point is that the exact locality where these events occurred is of paramount importance; this sense of locality is what ties Indigenous people to their local community in both the social and ecological sense.

Indigenous people view these connections as being very fluid. Any factor that alters a system, including tampering by humans, causes changes in many unpredictable ways. As Turner indicates, each species is constantly fluctuating, both in behavior and numbers, in response to many other species and to physical factors in the environment. An Osage scholar states: "The cosmos was in constant motion and consisted of unending, varied cycles of birth, maturity, old age, death, and rebirth. These temporal cycles could not be stopped or reversed, for 'nothing in the cosmos moved backward'" (LaFlesche 1995: 30). In modern ecological thought, physical factors are assumed to be directly connected to the living system, and organisms may influence the physical factors in their immediate vicinity (Jones 2005; Wright and Jones 2006).

When one species is removed, humans may notice its absence. More important, other species in the community are certain to notice this absence, because it probably alters their behavior or other ecological relationships in some way or another. One possible conclusion of this line of thinking is that Indigenous experiences are also shared by nonhumans. Indigenous people felt strongly about their involvement in the natural world, and at the same time they felt that they were not fundamentally different from any other species of animal (Deloria 1990; LaFlesche 1995; Pierotti and Wildcat 2000). They recognized that animals also try to understand humans and the rest of the world around them (see also Pennisi 2006). Humans are connected to all other beings, and this connection arises in large part from the physical proximity of all of these interacting parts.

HUMANS AS PART OF THE NATURAL WORLD

The worldviews and cultures of Indigenous American peoples evolved in the environments of the continents of North and South America, which means that these peoples came to depend upon the animals and plants of these environments for food, clothing, shelter, and, perhaps most importantly, social companionship. Identification with local plants and animals led to the development of strong ties to these nonhuman lives. "Little emphasized, but equally as important for the formation of [Native] personality was the group of other forms of life which had come down over the centuries as part of the larger family" (Deloria 1990).

The scholar Eugene Anderson writes in a similar vein concerning his experience with the Yucatec Maya of southern Mexico:

> The most striking characteristic of the technological system is that it maximizes total use of the environment. Everything is used, for as many purposes as possible, and as efficiently as can reasonably be done. . . . The technological system depends on a superb, locally fine-tuned knowledge base. . . . I have recorded about 700 species of plants and well over 200 animals known and named in Chunhuhub. Most are used for one or another purpose. Over 150 plant species are used for food, while almost 100 are used for poles or timber. About 300 are used medicinally. Interestingly, about 150 species are grown strictly as ornamentals, while the flowers of countless additional species are enjoyed in the wild; the Maya are not mere dusty economic rationalists, and aesthetics is a major consideration in their plant management. Similarly, almost any mammal or bird that will survive in the household may be kept as a pet; a compound often has many pet rabbits, guinea pigs, and parrots, and may even have a deer or peccary. Some species have a dozen uses. In addition, Maya farmers know many soil types, and how much of what kind of soil a given plant needs. They know when plants flower and bear fruit. They understand companion planting and biocontrol. They know virtually everything about the behavior and reproduction of wild animals, even obscure birds of no economic use. (Anderson 2005)

The body of knowledge acquired through this connection to local nonhumans and careful observation of these other species came to constitute much of what Indigenous Americans regard as TEK (Traditional Ecological Knowledge). As mentioned in Chapter 1, one major theme that emerges from this knowledge is the concept that *all things are connected*. The Indigenous knowledge base emerges from an association of interacting populations, which relates conceptually to the Western definition of an ecological community as defined above. It is clear that humans are an integral part of such a community and that interactions between humans and nonhumans can be as significant as those among human beings.

An important example of this type of thinking can be seen in the discussion about re-introducing wolves into Yellowstone National Park. The argument made by wildlife biologists and conservationists of European heritage to justify the re-introduction was that of all the species that inhabited Yellowstone when it was first made a park, all but one can still be found living in the park today. The missing species was the gray wolf. This is simply not true; human beings were also regular inhabitants and participants. Shoshone, Arapaho, and other human groups were an important part of that ecosystem.

The reason that Western conservationists do not consider humans to be a missing component of the Yellowstone ecosystem is that Western thought persists in defining "wilderness" as ecosystems without humans present (Gomez-Pompa and Kaus 1992). In fact, it is true of much of Western ecology that its practitioners consider systems where humans are present to be "disturbed," rather than "natural." Humans are regularly removed from areas designated as national parks, forest reserves, and wildlife areas (Dowie 2005). This is not true only in North America. For example, in South Africa and Namibia, Indigenous peoples, such as the !Kung and Juwasi Bushmen, were removed from national parks such as Etosha in Namibia, which had a profound impact on the behavior and ecology of lions, *Panthera leo*, that inhabited the park and had established a long-term symbiotic relationship with the human inhabitants (Marshall-Thomas 1994). In contrast, in Nepal and Australia, a more enlightened approach, co-management of national parks, has been established that allows Indigenous peoples to continue to inhabit their traditional homelands (De Lacy and Lawson 1997; Stevens 1997; Howitt 2001).

One reason that the role of Indigenous people as part of their ecological communities is so important is that they do not think of the nonhuman elements of their community as constituting "nature" or as "wilderness," but as part of their social environment (Standing Bear 1978; Allen 1986). Native Americans who adhere to this philosophy do not think of leaving a "house" to "go into nature," but instead feel that when they leave their shelter and encounter nonhumans and natural physical features that they are just moving into other parts of their home. "What we call nature is conceived by Native peoples as an extension of biological man, therefore a [Native] never feels 'surrounded by nature.' A [Native] walking in the forest . . . is not in nature, but is entirely surrounded by cultural meanings his tradition has given to his external surroundings" (Reichel-Dolmatoff 1996: 8–9).

The implications of such relationships are profound. Nonhuman elements are incorporated into the ritual representation of the community, which helps to establish a nature-centered belief system (Martinez 1994; Anderson 1996; Basso 1996; Owens 1998). This is the case, for example, among the Mayan people:

> Another component of this ideology is a sense that the forest is something that we humans must care for. Maya woodsmen are constantly trimming vines that overgrow trails, cutting little firebreaks around valuable wild trees, and logging or otherwise taking useful plants in a pattern deliberately calculated to thin the forest and weed out low-value plants. There is no "wilderness" here; the whole forest is known to be a mosaic of regrowing milpas, and is considered to be part of the cultivation system. It is thus managed as a garden, rather than neglected, fought against, or "protected." The Yucatec Maya cannot relate to the destructive pattern of use that they see as typical of other ethnic groups (both Mayan and Hispanic), but they also cannot relate to the absolute preservation of wild land that has come in with the international conservation movement. They simply can't understand it. They enthusiastically accept the goal of conserving land and wildlife, but they are aware that their traditional, unobtrusive use of the forest actually increases its value to wildlife. For one thing, they know which trees provide good wildlife food, and protect these. For another, their milpa cutting and firebreak development stops wildfires from spreading far, but does prevent the fires that open the forest and allow healthy regeneration. (Anderson 2005)

THE DEFINITION OF "PERSON" AND COMMUNITY MEMBERSHIP

The linked concepts of the importance of the local place in determining traditions, connectedness to nonhumans, and the idea of nature as home rather than as "other" have profound implications for Native conceptions of politics and ethics. Unlike dominant Western political and ethical paradigms that find knowledge of how human beings ought to act imbedded in the life of one's social (i.e., human) relationships, Indigenous Americans found within their concept of community instructions concerning how a person should behave as a member of a community consisting of many nonhuman persons—four-leggeds, winged ones, plants, and even landforms (Druke 1980; Deloria 1990, 1992, 1999a, 1999b; Pierotti and Wildcat 1997b, 2000, 2001; Tinker 2004). "The primary focus of creation stories of many tribes placed human beings as among the last creatures who were created and the youngest of the living families. We were given the ability to do many things but not specific wisdom about the world. So our job was to learn from other beings and to pattern ourselves after their behavior. We were to gather knowledge, not disperse it" (Deloria 1999b: 224).

In contrast to the ideas just expressed, Western thought has traditionally followed the lead of Aristotle, who defined politics and ethics as exclusively human realms. Values, ethics, and politics exclude all entities but other human beings. Therefore, respect and concern for their welfare are not owed to nonhumans and landforms. By Indigenous standards, Aristotle's notion of community membership was overly limited—in Indigenous communities politics and ethics are not limited only to human beings (C. Martin 1978; Allen 1986; Deloria 1990; Salmon 2000).

The inclusion of other living beings and natural objects into the category of "persons," which includes human beings, requires development of concepts of politics and ethics that include these other community members (C. Martin 1999; Pierotti and Wildcat 2000). The line between human and animals is so lightly drawn in American Indian cultures that it ceases to exist at certain points (Bruchac 2003). Among Ojibwa people, "The concept of 'person' is not synonymous with human beings but transcends it" (Hallowell 1960:21). Consideration of nonhuman entities, including landforms, plants, and animals, as individual persons who are part of their communities keeps humans attending to the specific entity and its particular good (see Taylor 1986 for a Western version of this concept). Such beliefs lead to what has been described as "kincentric" ecology, in which humans and nonhumans are viewed as part of an ecological assemblage that is treated as an extended family who share ancestry and origins (Salmon 2000).

Another illustration of how Indigenous peoples include other living beings as members of their community can be observed in clan names and totems, which reflect the existence of covenants between certain human families and specific animals (Deloria 1990; Pierotti and Wildcat 1997a). *Totem* is derived from the Anishinaabe word *ototeman*, which translates roughly as "my relative" (Bruchac 2003, 160). These nonhumans are connected to families over prolonged periods of time and offer their assistance and guidance during each generation of humans (C. Martin 1999; Pierotti and Wildcat 2000). If you have a certain creature as a totem you are not allowed to hunt or kill it (Bruchac 2003; see also Chapter 6), which may explain why in some cultures, such as the Pacific Northwest, only predatory species are used as clan signifiers. Throughout Native American cultures, there is a broad commonality of beliefs about animals in which human and nonhuman are bonded closely and part of one community involved with one another in terms of empowerment and emotional interactions (Deloria 1990; Martinez 1994; Anderson 1996; C. Martin 1999; Barsh 2000).

Links established through clan membership are based on the assumption that nonhumans are intelligent and can think about what they are doing and also what others are thinking. "To treat animals with 'respect' is to conduct one's relationship with them in accordance with the social standards that the animals deem appropriate and this may vary among species" (Nadasdy 2003: 93). As I mentioned in the previous chapter, humans work to establish covenants with other forms of life that serve to reduce the negative impact of human actions, which serve as the equivalent of "resource management" but among entities considered to be related and expected to interact as equals.

Assuming that the animals are sentient is linked to the concept referred to in Western science as a "theory of mind" (Pennisi 2006). Until the last few years Western science did not accept that nonhumans could have a theory of mind; however, recent discoveries have changed the way nonhuman thinking is perceived (de Waal 1995, 2001; de Waal and Tyack 2003; Bekoff 2002, 2006; Pennisi 2006). Most of this research involves nonhuman primates, such as chimpanzees, gorillas, and various species of monkey, but other studies have revealed similar attributes in birds such as Scrub Jays (*Aphelocoma coerulescens*), crows (*Corvus brachyrhynchus*), and ravens (*Corvus corax*) (Heinrich 2000; Emery and Clayton 2004), and in canids, such as foxes (*Vulpes*) and wolves (*Canis lupus*, which includes domestic dogs) (Hare and Tomasello 2005). These new developments reveal that Western science has had to rediscover knowledge assumed to be part of the understanding of Indigenous societies.

Perhaps even more indicative of the convergence of Western science and Indigenous knowledge is that over the last decade Western scientists have begun to realize that cooperation is more important in both nonhuman and human societies than is competition, which has been the basis of ecological thinking in the Western tradition (Bekoff 2007). Among nonhuman species that have been studied, it turns out that 85 to 95 percent of behavior is affiliative or cooperative rather than aggressive or competitive. Cooperation and "facilitation" are being found not only within species but among species as well, which is strongly tied to the Indigenous emphasis on coevolution and relationships (Barsh 2000; Shouse 2003).

Relationships implied by clan membership are more profound than most people can imagine, and the implications of such relationships carry consequences that might make adherents to the dominant culture uneasy or uncomfortable. To be a member of Eagle, Wolf, Bear, Deer, or even Wasp or Crayfish clan means that you are kin to these other persons; they are your relations. Ecological connectedness is culturally and ceremonially acknowledged through clan names, totems, and ceremonies (C. Martin 1999; Pierotti and Wildcat 1997a, 1999b). In nearly all Native American stories it is established that animal- and plant-persons existed before human-persons (Pierotti and Wildcat 1997a; Deloria 1999b). Thus, these kin have the role of elders, and much as do human elders, they function as teachers and respected members of the community (see quote from Deloria 1999b above).

Another recent discovery in Western science is that nonhumans as diverse as elephants, meerkats (*Suricata suricatta*), sperm whales (*Physeter catodon*), killer whales (*Orcinus orca*), and Western Gulls (*Larus occidentalis*) all show cultural transmission and teaching (Vogel 1998; Whitehead 1998, 2003a, 2003b; Annett and Pierotti 1999; Payne 2000, 2003; McComb et al. 2001; Clode 2002; DeWaal and Tyack 2003; Thornton and McAuliffe 2006). "Animals teach us things too . . . patience and respect and stuff like that . . . I always think of what we as a whole society can learn from animals . . . I mean we once lived with them" (Kluane First Nation elder quoted in Nadasdy 2003: 101). Acknowledging that nonhumans can function both as teachers and as elders requires that their lives merit special attention and consideration. It is also crucial to recognize that the lives of these nonhuman persons have meaning on their own terms and do not exist solely in terms of their utility to humans (Pierotti and Wildcat 1997a, 1997b, 2000).

Recognition of the value and meaning of nonhuman lives extends the social world to include nonhumans and generates an ethical system that requires their proper treatment. Humans live in mutual aid relationships with nonhumans (Barsh 2000). If humans eat or otherwise use nonhumans, they are empowered by that relationship. This empowerment leads to mutual respect. Many nonhumans have powers far beyond the capabilities of ordinary humans and are able to move with ease through worlds impassable to humans (Anderson 1996). "In his relations with animals the Yukon Indian meets beings with power which often seems to be greater than his own. If he is to live at all, he must find ways of coming to terms with these powerful co-inhabitants of the earth . . . animals are endowed with spirits and are potential givers of good . . . they can never be ignored" (McClellan 1975: 91). For example, birds move through the air at their own initiative, which is off limits to humans, and fish and marine mammals move through water in a manner that humans can only imitate in a clumsy fashion.

It follows logically that in Indigenous thinking that if nonhumans are "persons," they must have cognitive abilities, which would mean that they should recognize the danger of being hunted by humans. Thus if a nonhuman is caught, it was assumed to involve some element of choice on their part (Anderson 1996). This led to the concept of the prey "giving itself to you," which was the basis of the covenant that if you mistreat an entity, that entity may respond in a way that can cause harm to you and your people (see Chapter 3). This presumed gift required gratitude (thanks), as well as respectful treatment of the body of the nonhuman on the part of the human taking its life (Tanner 1979; C. Martin 1999). Although the prey may not truly give up its life voluntarily, this assumption is an important guiding principle of the rituals that ensured that hunters and fishers treat their take with respect, so as not to offend the prey (Pierotti and Wildcat 1999a). "If we do not show respect to the bear when we kill him, he will not return" (traditional Mistassini Cree, cited in Bruchac 2003: 155). As Darwin's defender Thomas Huxley said of the tendency towards selfishness in human social behavior, "Let us understand . . . that the ethical process of society depends, not on imitating the cosmic (natural) process, still less in running away from it, but in combating it" (Huxley 1894). I would substitute "coming to terms with" for "combating," but otherwise Huxley's rule of dealing respectfully with others applies as much to nonhumans as to humans.

THE CONCEPT OF "BALANCE"

As described in Chapter 1, a key point to understanding TEK is to realize that Indigenous Americans integrated spiritual and ecological knowledge and understanding, blending these into a traditional way of life that allowed people to survive over extended periods of time under ecological conditions that were continuously fluctuating and relatively unpredictable. They managed to accomplish this without the use of pesticides, herds of domestic animals, or large-scale agriculture.

Surviving in a fluctuating environment is a very different reality than living in "harmony" with the "balance of nature," as Indigenous life is often caricatured (Ross 1992; Pierotti and Wildcat 1997b, 1999b). The assumption of a "balance of nature" whereby ecosystems, ecological communities, and animal populations are presumed to exist under equilibrium conditions to which they are inclined to return any time they experience a perturbation is a Western concept, which has its roots in the attempts of European philosophers to link science and religion through economic metaphors during the seventeenth century. "The creator had designed an integrated order in nature that functioned like a single, universal, well-oiled machine" (Worster 1994: 39). In the scientific literature statements on this concept can be found in Darwin (1859): "the forces are so nicely balanced that the face of nature remains uniform, for some time." A contemporary of Darwin's phrased it thus, "Nature, left undisturbed, so fashions her territory as to give it almost *unchanging permanence of form, outline and proportion*, except when shattered by geological convulsions . . . in these comparatively *rare cases of derangement* she sets herself at once to repair the superficial damage and to restore, as nearly as practicable, the former aspect of her dominion (Marsh 1864: 29, emphasis added). Such concepts continue into recent science as envisioned by European investigators: "There is a balance, there is an almost permanence about such animal communities. We see the stability, the lack of change" (Kruuk 2002).

This concept of balance and associated stability underlies all of the classic ecological models in population biology. These ideas were codified and given mathematical form by the Lotka-Volterra equations developed in the 1920s (Kingsland 1985). Ecologists from the Western tradition came to take literally their metaphor of "balance." They considered this an accurate representation of how ecosystems functioned, and their research questions were based upon the assumption that "balance" was an actual state that ecosystems strove to attain. Associated with this metaphor was the assumption of the literal truth of the idea of "equilibrium" communities and "*K*-selected populations" that reach a "carrying capacity" at which population numbers remain relatively constant. Upon reaching carrying capacity, population size is maintained almost exclusively through density-dependent processes resulting from intraspecific competition (Lack 1954; Levins 1968; MacArthur 1972; May 1973; see Ricklefs 1998 and Kingsland 1985, 2005, for summaries).

The metaphor of balance and equilibrium led inexorably to one of the most pernicious of all models for the exploitation of natural populations of birds, fish, and mammals, the idea that populations can exist at a level that allows a maximum sustainable yield (MSY). MSY assumes that it is possible to exploit a population when it is near one-half of its carrying capacity (*K*). If an exploited population is assumed to be near the MSY point, it can be exploited without limits. Unfortunately under fluctuating environmental conditions *K* is hard to define and probably does not exist in any meaningful sense (Larkin 1977; Pierotti 1991a). Nonetheless, the MSY concept became the fundamental model for determining appropriate levels of "exploitation" in wildlife and fisheries biology (May 1973; Holt 1975; Larkin 1977).

The MSY concept also employs the "simplifying assumption" resulting from the idea of a carrying capacity (defined as *K*) that population sizes of fish and wildlife are regulated almost exclusively by competition within a species, and that predation and environmental fluctuations are of little consequence (Larkin 1977). Application of such models has led to disastrous crashes of populations of large fish throughout the world, including cod, *Gadus morrhua*, in the North Atlantic (Kurlansky 1997; Myers and Worm 2003, 2005).

Another simplifying assumption of MSY and equilibrium models is that all individuals in a population are equivalent and can be substituted for one another. One implication of such thinking is that newborn individuals are functionally equivalent to breeding adults (see also Chapter 4). Clearly the scientists who employ such models realize that this is not true, but they continue to employ such models to design management schemes for natural populations. Largely ignored in these simplified models that dominate modern fish and wildlife management, as well as much of population ecology, is any understanding of the dynamics and functional structure of actual animal populations. Some Western scientists have suggested alternative models that are much more realistic (e.g., Holgate 1967; Murphy 1968; Mertz 1971a, 1971b), but these alternatives were pushed aside in the quest for balance-based, equilibrium explanations, for example, the mania for "*r* and *K* selection" that swept population ecology in the 1960s and 1970s (Pianka 1970; MacArthur 1972; Boyce 1984).

DEALING WITH A CHANGEABLE AND UNPREDICTABLE ENVIRONMENT

Indigenous peoples depended upon their nonhuman relatives for food and shelter; therefore these peoples would be aware of the actual conditions existing in local populations of four-leggeds, winged ones, swimmers, and plant persons and how environmental changes could also impact these relatives. Indigenous traditional beliefs and TEK, along with the accompanying rituals and ceremonies, appear to have been established for three purposes: (1) to minimize negative impacts on species upon which human populations depended for food and shelter, especially during periods of slow decline, which probably represent the majority of the time; (2) to placate and protect the individuals who are most likely to survive the lean conditions and propagate their kind for the future; and (3) to make sure that the people are prepared and in position to take advantage of boom conditions while remaining cognizant of the role of environmental unpredictability on their nonhuman relatives and the potential impacts upon their own future survival and that of their kind (see also Ross 1992).

Indigenous Americans were aware of the true nature of population and environmental fluctuations because they kept constant track of the changeable non-equilibrium conditions that predominate in the real world. During the period when modern human beings were evolving over the last 100,000 years there have been only two generally stable periods of climate (Pearce 2007: 237). The first was when the ice sheets were largest and the world was coldest, and the second is the period in which we are living now. Ironically, the Western scientific tradition has treated this most recent period as if it were typical. Thus, it has only been in the last 25 years that Western ecologists have begun to recognize the changeability of the natural world and to reject their older models based on the metaphors of "balance of nature" and "equilibrium" communities and populations (Botkin 1990, 1991; Hoffman and Parsons 1997; Coates 1998: 186–91; Pearce 2007). Despite this realization, these updated ideas have yet to make it into most ecology textbooks.

In contrast, ceremonies and stories of Indigenous Americans emphasize the changeable and unpredictable nature of the environment. One classic story is reported from the Lakota of a summer when the rain would not stop and the waters rose until the land was flooded. "Out of the north came the winds. . . . They whipped the floods into a mean-spirited dark being. . . . Within days all but one of the people was dead" (J. Marshall 2001: 126). This one woman survives to become the progenitor of the Lakota people.

Compare this story to the description of the end of the last Ice Age: "In the final millennia as melting made fitful but sometimes dramatic progress, a very large amount of liquid water was produced. Often it did not pour directly into the ocean, but formed giant lakes . . . on the land around the edges. The largest of these is called Lake Agassiz. . . . It stretched more than

600 miles across the American Midwest from Saskatchewan to Ontario and from the Dakotas to Minnesota, generally moving with the advancing front of warming. . . . In the early stages of glaciation the lake drained south, down the Mississippi River. . . . But about 12,800 years ago . . . something stopped this and forced the lake to drain east" (Pearce 2007: 151).

This change in Lake Agassiz would have sent it directly over the lands in which the ancestors of today's Lakota lived. The description of the weather in the story reported by Marshall is essentially what we would expect to happen as massive glaciers melted quickly, something that may have happened in as little as a decade (Pearce 2007). It is likely that what Marshall (2001) describes as a re-creation story of his people is also an account of the end of the Ice Age and its consequences to his ancestors.

This recognition of a changeable and unpredictable nature, combined with the realization that such things could happen again, underlies the rituals involved in giving thanks to animals and plants after taking them for human use (Pierotti and Wildcat 1999a; Pierotti 2005a), as well as ceremonies such as First Salmon ceremonies in the Pacific Northwest (Raban 1999) and the Sun Dance of the Plains Indian tribes (Harrod 2000). Despite numerous references to "keeping things in balance" (e.g., Krech 1999a), Indigenous ceremonies and rituals were based on an understanding of non-equilibrium population dynamics and a realization that the natural world was almost never "in balance" in the sense of remaining constant and unchanging. "The ceremonies themselves become ways of coming to know, of understanding" (Cajete 2000: 81).

It is in the best interest of human societies to try and minimize risk when dealing with key food supplies. This is one reason that gathering by women is probably more important than hunting by men in maintaining the basic sustenance of many Indigenous peoples around the world. This also explains why, during the majority of times, little or nothing is wasted during hunting activities (Tanner 1979; Pierotti 2005a; Henriksen 2009). Rituals associated with minimizing waste are therefore codified as "religious or spiritual" (Tanner 1979; Brody 1982, 2000; Anderson 1996; Pierotti 2005a; Henriksen 2009), because caring about your prey makes you much more likely not to harm it and reduce its value as a resource (Anderson 1996).

The attitudes and relationships of Indigenous people with regard to other organisms result from having evolved as distinct cultures in strong association with those other creatures and experiencing them on a daily basis (C. Martin 1978, 1999; Pierotti and Wildcat 2000; Nadasdy 2003). To Indigenous peoples, relationships with the nonhuman and human world are the basis of their religion or spiritual belief system and also of a way of life (Nadasdy 2003). Thus religion serves to code knowledge about community dynamics (Rappaport 1971; Deloria 1992; Anderson 1996). Such forms of religion provide direct emotional involvement with the nonhuman world and thus avoid the Romantic need for transcendence.

For example, Laguna Pueblo people could not have survived in the arid southwestern United States without their recognition that they were "sisters and brothers to badger, antelope, clay, yucca, and sun" (Silko 1996). "Kluane people's knowledge of the land and animals was necessarily vast for, without such knowledge . . . they would have long ago perished" (Nadasdy 2003: 67). To Northwest Coast peoples, "fish, bears, wolves, and eagles were part of the kinship system, part of the community, part of the family structure. Modern urbanite ecologists see these as Other, and romanticize them, but for a Northwest Coast Indian, an alien human was more Other than a local octopus or wolf" (Anderson 1996: 66). The Raramuri (Tarahumara) people of northern Mexico use the term *iwigara* to indicate the way in which they are bound to the land, animals, and winds of their Sierra Madre home. *Iwigara* indicates the interconnectedness and integration of all life in the Sierra Madres, both physical and spiritual (Salmon 2000).

Adherents to this concept of community recognize that nonhumans existed before humans did. To Silko's Laguna people, emergence into this world was possible because of help from badger and antelope. In Numic (Comanche or *Nuhmuhnuh* and Shoshone or *Newa-nuh*) tradition, wolf was the creator who taught humans how to hunt and how to live in the world (Ramsey 1977; Buller 1983). In the Rock Cree, animals were recognized to have existed before human beings, and humans were known to come from animals during the progression of the earth (Brightman 1993). The Oglala Lakota believe that wolves were a nation long before human beings considered themselves a people (Iron Cloud cited in McIntyre 1995).

Recognition of such connections between humans and nonhumans leads also to the concept that *all things are related*, which is central to Indigenous spiritual beliefs (see Chapters 4 and 5). Related concepts in Western science are less than 150 years old. Darwin's (1871) demonstration that humans must have evolved from nonhuman ancestors was considered to be a revolutionary concept because it ran counter to prevailing Western philosophy, following the Aristotelian tradition, which separated humans from the rest of nature. Perhaps the most important consequence of Charles Darwin's theory of common descent was its change in the position of humans from separate from nature to part of nature (Mayr 1997). This theory served to establish in Western thought the tenet that humans are related to nonhumans and linked to them through community dynamics.

Often unrecognized is that not only are humans dependent upon the nonhuman, but also that the reverse is often true. Activities of humans are often important in shaping the lives and ecology of the nonhuman elements of their communities. Burning practices of Indigenous peoples of both North America and Australia are known to have major effects on local plant community structure and lead both to increased biodiversity and increased population size of many important species (H. Lewis 1989; Boyd 1999; Williams 2002; Levy 2005). Bison depended upon humans who burned the prairie, especially the tall grass prairies, to maintain an ecosystem that allowed these large social ungulates to exist in such great numbers (Isenberg 2000). In contrast, until recently both Western science and popular culture have considered "wildfires" to be both "highly disruptive and environmentally destructive." Only very recently has Western science come to realize the value of fire as an important component structuring ecological communities (Boyd 1999; Levy 2005; Wuerthner 2006).

FOOD PRODUCTION AND WORLDVIEW

It is fairly obvious that Eurasia and America went down quite separate paths with regard to perceptions of both the meaning of community and the natural world as a whole. This emerged from the way these traditions tried to create reliable sources of animal protein. Eurasia (and Africa) turned to domestication of animals, especially social ungulates (cattle, sheep, goats, pigs, and horses). As a result the Eurasian tradition either failed to develop or refused to accept the concept of nonhuman animals as persons, because they devalued the life of their ungulate *chattel* (an old French word that is derived from the word *cattle*). This also led to the introduction of a wide range of animal-originated pathogenic organisms that jumped from their original ungulate (and bird) hosts (see below and Pierotti 2004).

In contrast, Indigenous North Americans had no domestic animals except wolves (dogs). Even in Central and South Americas there were few domestic animals because the Americas had few species readily susceptible to domestication (Diamond 1997). Instead Indigenous Americans relied on their ceremonies and profound knowledge of the animal, bird, and fish populations upon whom they depended to minimize chances of running out of important sources of food.

In Europe and Asia domestication of animals probably predated the domestication of plants. Human hunters typically wandered with herds. In Eurasia the herding ungulates whom hunters followed had social behavior that made them susceptible to domestication—they lived in herds with well-developed dominance hierarchies and occupied overlapping home ranges rather than territories (Diamond 1997: 197).

Domestication of plants is a different process and took place in both the Americas and Eurasia. Indigenous Americans lived and worked with the Three Sisters: corn, beans, and squash (Mt. Pleasant 2001, 2006). These plants are referred to as "sisters" because of the way they interact. The corn acts as a support pole on which the beans grow. Beans, which are legumes, fix nitrogen in the soil, thus providing nutrition for both corn and squash. The growth form of the squash as a widespread vine, combined with its hairy leaves, prevents herbivores ranging from insects to deer from getting to the beans and corn. Thus you have three species, all of which enhance the growth and production of the other two (Bruchac 2003; Mann 2005). Indians also grew peppers, peanuts, and root crops such as potatoes, and altered habitats to maximize production of berries, fruits, and nuts. Indigenous American hunter-gatherers also carried bulbs and seeds of important food plants that were not formally domesticated, such as prairie turnips, and planted these in suitable habitats so they would have a wider distribution of these important food sources. In contrast to Indigenous American food plants, Eurasian cereal grains (wheat, barley, rye, rice) are much less productive than corn and much less rich in protein than beans (Mann 2005). Europeans compensated for this reduced nutritional value through their use of the meat of their domestic mammals, whereas Indigenous Americans supplemented their plant diets with animals and fish taken as prey (see Chapter 3).

Domestication of plants does not appear to lead to a lack of respect for nonhumans. This is probably because plants, especially when grown in a polyculture such as the Three Sisters, seem to retain their essential nature as well as their ecological relationships to one another. In contrast, Eurasian morals and ethics evolved to minimize the recognition of relatedness between humans and other animals. It is difficult to treat relatives as chattel or moveable property that you control. In fact, even when contemporary scholars attempt to establish the importance of recognizing the "good" of nonhuman lives, they are forced to develop lengthy and complex arguments to establish the idea that nonhumans should be included as community members (e.g., Taylor 1986). Animal Rights activists struggle with this issue, as do evolutionary biologists. The European philosophical and religious tradition has difficulty with the twinned concepts of relatedness between humans and nonhumans and respect for nature, which are fundamental concepts in Indigenous spirituality (Scott 1997; Deloria 1999b).

The Western cultural tradition made a major effort to separate itself from any association with the natural world, except as a source of resources for exploitation (Sale 1991; Coates 1998). In particular, various sects of Christianity developed a philosophical tradition that "offered no encouragement for any investigation into the foreordained ways of God's creatures, much less the established workings of his trees and rivers and soils . . . it was sufficient for them to know that God created them, blessed them, and then gave humans 'dominion' over them" (Sale 1991). This attitude was the result of a long tradition that felt "Earth is full of restless dread throughout her woods, her mighty mountains, and deep forests" (Nicholson 1959; Sale 1991). In seventeenth century England, "the encroachment of wild creatures into the human domain was always alarming" (Thomas 1996).

Mountains were places of dread, "regarded as physically unattractive," and "early modern travelers found mountainous country unpleasant and dangerous" (Nicholson 1959; Sale 1991). Forests were even worse, because of an imagined bestiary (including werewolves, vampires, and Pan, the goat-legged god of the woods and source of the word *panic*) that inhabited these areas, "but forests and mountains need not be populated to be fearsome. It was enough that these places were wild: that was the trigger to the terror" (Sale 1991). It became part of the Romantic Tradition to seek out experiences that left the heart in a state of terror because this left them with a more intense experience as they sought the "sublime" (Worster 1994, Chapter 6). This tradition continues to this day in experiences as diverse as horror films and mountain climbing.

The wild was "so unreasonably fearsome that the encroachment of wild creatures into the human domain was always alarming" (Sale 1991). A bee flying into a cottage or a bird rapping at the window was enough to send strong men to bed, and in 1604 the English House of Commons rejected a bill in 1604 after the speech of its sponsor had been interrupted by the flight of a jackdaw flying through the chamber, "an indisputedly bad omen" (Sale 1991; Thomas 1996).

Such traditions originated in the Renaissance of the fourteenth and fifteenth centuries, which emphasized absolute human autonomy (Coates 1998). The seventeenth-century scientific revolution did little to change this scenario; in fact, it actually made the situation worse by "transforming nature from a living organism into a machine—simple, unfeeling, inert matter with no intelligence, soul, or purpose—the new mechanistic philosophy assisted the commodification of nature" (Worster 1994; Coates 1998: 68). The eighteenth-century "Enlightenment" stressed that humans were masters of their own destinies and emphasized the subjugation of nature (Coates 1998). The Europeans who emigrated to North America during the seventeenth and eighteenth centuries were disciples of this cultural, philosophical, and intellectual tradition.

Given this tradition, it is not surprising that when Europeans came to North America, they regarded the "wilderness" as threatening and hostile. Even the earliest explorers regarded America as a land full of uncontrolled and frightening peoples and animals (Martin 1999). The Reverend David Jones, who spent time living with the Shawnee people in Ohio in the 1770s, described the experience as "like living with lions" (Gilbert 1989), which is quite a compliment when you really think about it, although it is doubtful that the good reverend intended it that way. Once Europeans learned of the philosophical and spiritual traditions of the Indigenous peoples, they felt compelled to regard these beliefs as "primitive and savage"; after all, these belief systems emphasized ties to nature or the wild that filled Europeans with fear (Martin 1999). Since the time of European contact, especially from the late eighteenth through the early twentieth centuries, there have been consistent attempts by the dominant culture to destroy the notion of ecological communities as extended families.

CONSEQUENCES OF COLONIZATION

It is instructive to examine the ecological consequences of the interaction of European and Indigenous American societies that began with the "discovery" of America by Europeans, including Columbus, Cortez, Pizzaro, and Cartier. There appears to have been a dramatic transformation in the behavior and attitudes of Indigenous peoples of the Americas with regard to their relationship with the natural world, especially toward animals, that followed upon the arrival of Europeans during the seventeenth and eighteenth centuries (Martin 1978; Mann 2005).

This change in attitude seems to have been a result of several factors: (1) introduction of contagious diseases into human populations that had no evolved immune responses to these pathogens, (2) consequent social devastation resulting from epidemics that may have killed as much as 90 to 95 percent of affected populations (Martin 1978; Diamond 1997; C. Mann 2005), and (3) introduction of a market economy, which required production of products, predominantly furs or hides, that could be exchanged for trade goods, especially metal and cloth produced by Europeans. The exact dynamics and sequence of these events is hard to establish, but it is clear that one consequence of these events was overexploitation of fur-bearing mammals, such as beaver, and possible overhunting of bison and deer in some areas (Martin 1978; Krech 1999a; Kay and Simmons 2002). These activities, which appeared to lead to overexploitation of animal populations, have been used to criticize the ecological understanding of Indigenous peoples (Krech 1999a, 1999b; Kay and Simmons 2002).

Almost all contagious diseases that have caused major health problems and epidemics over human history are known to have originated in nonhuman animals and subsequently "jumped the species barrier" into humans (Martin 1978; Diamond 1997; Pierotti 2004). As a consequence, animal and human diseases cannot clearly be separated when discussing the impact of disease on human environments. Humans are mammals; therefore diseases found in nonhumans, especially other mammals, can often cross readily into humans. Prominent examples of such contagions include smallpox, cholera, tuberculosis, bubonic plague, various forms of influenza, malaria, and AIDS (Cockburn 1967; Pierotti 2004; Cohen 2009).

Virtually all animal diseases that jump to humans as a new host are caused by bacteria and viruses whose small size and evolutionary history make them highly volatile and hence more likely to be transmitted from one individual to another, which is the basis of contagion (Cockburn 1967; Diamond 1997). These disease organisms have coevolved through interactions with other species, which have also evolved an immune response to them so they are not generally serious health problems in their original host species. What makes most of these diseases so virulent in human populations is that when first exposed to them, humans have no evolved immune response to these pathogens. For example, smallpox is related to bovine pox, which causes minor problems in cattle but is often fatal in its mutated form in humans (Cockburn 1967). Similarly, the AIDS virus is closely related to a viral infection that occurs in African primates, where it causes only mild influenza-like symptoms (Pierotti 2004). Other examples include

(1) measles, which is closely related to the ungulate disease rinderpest; (2) tuberculosis, which is closely related to a similar disease in cattle; (3) and influenza, which is actually a complex of viral diseases derived repeatedly from similar pathogens occurring both in pigs (swine flu) and birds such as ducks and chickens (both the major influenza epidemic around the time of World War I and the current bird flu scare are the result of influenza viruses that originated in birds).

The domestication of ungulates such as cattle and swine set up a scenario whereby humans in Europe and Africa were continually exposed to a wide range of epidemic diseases. These diseases thrived within the high densities at which human societies kept cattle and pigs (Pierotti 2004). European farmers are traditionally sedentary. In historical times farm families were in close contact with their own sewage and that of the domestic animals with whom they lived in an intimate and symbiotic fashion. In many agrarian societies farmers used to take cattle and pigs into their homes at night, both for warmth and to protect their livestock from predators (Cockburn 1967), which both enhances and prolongs the exposure of the human population to bacterial and viral pathogens (Pierotti 2004).

Agriculture sustains much higher human densities than the hunting-gathering lifestyle that agriculture replaced (Diamond 1997). Thus, the large concentrations of humans resulting from increasing urbanization in Europe provided fertile ground for the rapid spread of infectious diseases. Only within the last century did European cities achieve self-sustaining populations, because so many city dwellers died from contagious disease that constant immigration from rural areas was required to sustain urban areas (Cockburn 1967). For example, bubonic plague is spread by fleas, which pick up the plague bacillus from the fur-bearing mammals that are their normal hosts. Plague first appeared in Europe in the mid-sixth century and had devastating impact in fourteenth-century continental Europe (called the Black Death), where it killed as many as 25 million people (Gottfried 1983). In the seventeenth century plague killed 25 to 30 percent of the total population of the British Isles (Gottfried 1983; Diamond 1997).

Some New World cultures existed at densities comparable to or greater than those found in Europe (Mann 2005). The Aztec capital of Tenochtitlan may have been one of the largest cities in the world during its heyday. One consequence of these dense American societies is that the residents, like Europeans, turned more toward hierarchical structures and concentrated wealth. The Aztecs even began using victims of human sacrifice as animal protein to supplement the vegetable protein in their diet (Harner 1977; Mann 2005).

These large, complex societies were also the only Indigenous American cultures to turn to domesticating animals, such as llamas and guinea pigs (Diamond 1997; Mann 2005). Unlike the domesticated ungulates of the Old World, these New World domesticates were never maintained at high densities, and humans did not drink their milk. None of these animals except dogs were kept in close proximity to humans, as livestock were in the Old World. This seemed to prevent diseases jumping from the animals to humans (Diamond 1997; Mann 2005). Even though dogs and humans share many physiological responses, humans that live with dogs are generally immune to life-threatening diseases specific to canids (Clutton-Brock 1984; Serpell 1995; Pierotti 2004).

In addition to the Aztecs, other New World communities such as the cities of the Mayans and Inkas, and the Mound Builder cultures along the Mississippi and Ohio River valleys lived at densities comparable to those found in European and Asian cultures at the same period in history. Epidemic diseases, however, appear to be virtually non-existent in these Indigenous New World cultures, which is almost certainly attributable to the absence of domestic ungulates who are known to have been the source of epidemic diseases in Europe, Asia, and Northern Africa (Cockburn 1967; Diamond 1997).

The absence of nonhuman-derived contagious diseases and associated immune responses in New World humans was certainly a major factor in the successful invasion of the New World by Europeans. Introduction of animal diseases into susceptible human populations represents what has been referred to as the first, or microbial, phase of the European conquest of the Americas (C. Martin 1978; Crosby 1972; Mann 2005). Despite the success of supposed conquistadors like Cortez and Pizarro, it was actually smallpox that led to the collapse of the Aztec and Inka empires (Mann 2005). Cortez's initial 1519 foray into the Aztec civilization was much less successful than his subsequent 1520 effort, which occurred after smallpox arrived in Tenochtitlan. By the early seventeenth century the Indigenous population of Mexico had fallen from an estimated 20 million to less than 2 million (Mann 2005; Pierotti 2004). In a similar fashion, smallpox arrived in Inka territory in 1526 and set up the opportunity for Pizarro's successful "invasion" in 1531. There are well-documented cases of 90 percent or more of Indigenous populations being wiped out by these new contagious diseases that arrived with Europeans and their symbiotic nonhumans. As one example, the Mandan, one of the most elaborate of the Great Plains cultures, suffered mortality of more than 95 percent of their population after arrival of smallpox on a Missouri riverboat in 1837 (Crosby 1972; Martin 1978).

Not only humans suffered as a result of the introduction of these diseases. It is rarely discussed, but many natural populations of animals, including deer, caribou, moose, bison, and beaver, also experienced massive die-offs from west of Hudson Bay to the Rocky Mountains during the latter part of the eighteenth century (Martin 1978). These deaths probably also resulted from disease introduced by Europeans through their domestic animals. These die-offs were almost exclusively among ungulate populations, who would have been most susceptible to the ungulate-derived contagious diseases characteristic of Europe and Asia (Cockburn 1967; Diamond 1997).

Mammalian carnivores appeared relatively unaffected by these pathogens, but would have suffered as a result of loss of their ungulate food sources.

Perhaps the major consequence of this introduction of nonhuman-derived diseases was the destruction of cultural traditions based on respect for nonhumans (see Martin 1978 for a thorough, detailed account and interpretation of these events). As emphasized above, most of the Indigenous cultures of North America had philosophical traditions where nonhumans were regarded as creator spirits, and the concept of relatedness was based upon ecological relationships. The devastating impact of introduced disease on these cultures may have caused them to turn on their nonhuman relatives, which led some tribes to be willing to wipe out local populations of beaver, deer, bison, and wolves in order to trade furs for European trade goods and metal (Martin 1978; Kay and Simmons 2002; Pierotti 2004).

Epidemic diseases not only devastated populations but also destroyed cultural traditions and the confidence established through these traditions. The impact of suddenly having 80 to 90 percent of a community or village killed by disease would introduce fear and a feeling of helplessness into peoples who had viewed themselves as strong, powerful members of their ecological communities. Calvin Martin (1978) argues convincingly that disease destroyed the assumed relationship between tribal communities and the "Keepers of the Game" who maintained and controlled animal populations (see Chapter 4 for a detailed discussion of this concept).

Because these diseases result primarily from close association with domestic animals, Indigenous American populations had no evolved immune responses (Diamond 1997; Mann 2005), whereas Europeans did. Therefore, even though individuals might sicken or even die while serving as vectors for these diseases, there were no epidemics among Europeans in the Americas. This supposed immunity, especially among clergy, suggested to surviving Indigenous Americans that people of European ancestry were protected by their "God" and were not the targets of the pestilence that was devastating their populations (Martin 1978).

Ironically, disease inflicted a similar impact on European society in the centuries prior to colonization of the Americas. One important, although often unappreciated, consequence of the fourteenth-century plague was its profound impact on European philosophy and science. The prevailing worldview in Europe prior to the mid-fourteenth century was mythic and symbolic, rooted in an idea of cyclical time, and placed far more emphasis on links between human and nonhuman aspects of the world than did the worldviews that arose after the Black Death (Gottfried 1983; Coates 1998; Pierotti 2004). When the plague arrived and began to have a devastating impact on local populations, the knowledge base and techniques of this older philosophical tradition were pressed into service, including prayer and medicine based on sympathetic magic. None of these methods proved effective, and the lack of ability to deal with the resulting death and devastation created both widespread panic and subsequent culture-wide depression (Gottfried 1983). The impact of massive, inexplicable loss of life on a society cannot be overestimated. The impact of the events of 9/11 on the U.S. society was huge, even though fewer than 3,000 people died as a result of those incidents, so it is difficult to imagine events that would have killed millions, as happened to Europeans during the Black Plague and to Indigenous Americans after the arrival of European invaders during the sixteenth and seventeenth centuries. Belief in spiritual traditions and ways of understanding how the world works were crushed in both Europeans and Indigenous Americans, which almost certainly led to a sense of spiritual desolation.

The experience of the plague, described by some historians as the "greatest biological-environmental event in history" and the "equivalent of Nuclear holocaust" by others, forced Western Europe to develop a new way of organizing its perception of reality (Cockburn 1967; Gottfried 1983; Diamond 1997). Within "Christian" societies the plague led to loss of faith in a benevolent, heedful creator followed by persecution and scapegoating of "heretics." This eventually led to the beginnings of Protestantism and its images of a vengeful, wrathful God (Gottfried 1983; Pierotti 2004).

From a more scholarly perspective, response to the plague experience led to development of the intellectual tradition that separated mind from body, objective from subjective, and human from nature. This led to the beginnings of the Renaissance and development of the Western European "rationalist" scientific tradition, ultimately generating Cartesian Dualism, the machine model/metaphor as a way of understanding nonhuman life, and the Baconian-Newtonian worldview (Sale 1991; Coates 1998; Pierotti 2004). Thus, the philosophical and spiritual impact of the plague led directly to the "modern" rationalist approach in which experimentation and measurement substituted for observation and experience.

This new way of dealing with reality had positive effects for Europeans, which is why it has come to dominate their way of thinking. It led to increased sanitation, which reduced the background levels of many contagious diseases. The division of reality into separate spheres of mind and matter provided a powerful methodology for the study and understanding of the "outside" world. It was largely inadequate, however, for understanding inner experience, the human mind, and our relationship with the world of our fellow life forms (Pierotti 2004). Thus, although this dualistic view led to increased sanitation, there was no increased understanding of the natural cycle of disease or the evolution of immune responses.

CONTINUING IMPACTS OF COLONIALISM

Subsequent to the impact of disease, the European invaders of "America" continued to attack the social structures and belief systems of Indigenous Americans. Much of this effort involved removing many peoples from the places where they developed and evolved their cultural traditions and belief systems through "relocation." Missionaries and other Christian proselytizers worked hard to convert "savages" away from their primitive belief systems and to indoctrinate these peoples into various Christian faiths (Martin 1978; Harrod 2000).

Another way in which tribal communities have been disrupted and their bonds to their nonhuman relatives have been broken is through the "education" system of the European tradition, especially the boarding school system, in which students were removed from their parents so that they could be forcibly educated in the traditions of the dominant culture. One of the saddest consequences of this experience is that the Indigenous peoples of the Americas, who may have been among the most bureaucracy-free societies in human history, have been turned in today's world, through the joint efforts of the Bureau of Indian Affairs, tribal governments, and their protectorate status, into some of the most bureaucratically constrained and limited groups on the planet. Today, even Indigenous identities are determined by bureaucratic processes.

Although the harmful aspects of this system of education have been extensively documented (Noriega 1992), it is rarely acknowledged that one major impact was the separation of Indigenous children from their connections with the natural world. Among the casualties was their relationship with the nonhuman aspects of their social systems, which had already undergone massive impact from overhunting and habitat destruction as Europeans invaded North America (Dunlap 1988; Harrod 2000).

Ironically, a major change in the attitudes of Indigenous Americans toward nonhumans may have resulted through introduction of Eurasian domestic mammals. The horse culture of the Great Plains tribes changed the attitude of these peoples toward animals, and they began to regard horses as "property" (the Indigenous American version of chattel). The large amounts of grazing area required by horses may also have had a negative impact on populations of bison, elk, and other Indigenous grazing ungulates and introduced a concept with which Indigenous cultures were previously unfamiliar— competition for grazing between domestic and wild animals (Isenberg 2000). Some tribes recognized at least some of the ethical issues involved. For example, in the Comanche people it was possible to obtain spirit power from all members of the nonhuman world, with two exceptions. These were the horse and the dog, who, having no control over their own lives, had no power to convey (Wallace and Hoebel 1948).

During the invasion of North America by Europeans Indigenous peoples, who were removed from their ancestral lands, experienced some loss of a sense of place, which has been reinforced by attempts at relocation, assimilation, and even termination (Churchill 1995). One striking consequence of this loss of connection can be illustrated in a study conducted among Salishan peoples in the early 1990s about the reintroduction of wolves into northern Washington. Older people over sixty years of age indicated fondness for wolves and felt a sense of connection to these predators. People between thirty and sixty indicated indifference but felt that wolves were potentially harmful. People under thirty, however, feared wolves and indicated that they disliked these animals and would shoot them on sight (Pierotti and Wildcat 1997b). This indicates the level of indoctrination and assimilation of Native youth, even in communities where elders are present and hold very different views.

For too long contemporary Indigenous Americans have allowed themselves to become obsessed with the values and attitudes of the dominant culture. It is clear, however, that today, largely as a result of its ecological ignorance, the dominant culture is in trouble and is justifiably feared by our nonhuman relatives. What is even sadder, however, is that our nonhuman relatives have learned to fear Indigenous peoples as well, because Indigenous people have learned to think and act like Europeans. For Native American peoples to truly restore themselves and their communities as anything more than pale copies of the dominant culture, it is clear that Indigenous people must re-establish meaningful contact with the nonhuman elements of our communities. This means developing and showing respect for the places in which they live and for the nonhumans. The nonhumans are still willing to share these places, as indicated by their desperate attempts to hang on despite hatred and persecution (Dunlap 1988; Marshall-Thomas 1994).

Environmental philosopher Wes Jackson has written a book-length essay entitled *Becoming Native to This Place* (Jackson 1994), which he describes as an "instruction manual" for people of European descent to establish ties and respect with nature, because he feels this is the only way to a viable future.

One of Jackson's most compelling stories in his book is his recounting of how he cleared brush behind his house. Later that evening Jackson was surprised to find a raccoon (*Procyon lotor*) fighting with his dogs in the backyard. Jackson wondered why this had never happened previously. It then occurred to him that earlier that day he had destroyed the habitat of the raccoon, and the raccoon and his dogs were now working out a new social order that was the direct result of his anthropocentric act. It would be ironic if one consequence of colonization for Indigenous Americans, who are truly "native to this place," was that they forget their relationships with and obligations to the other members of their communities at the same time Euro-Americans are striving to develop such relationships.

3 Predators Not Prey

"Wolves of Creation" Rather Than "Lambs of God"

> "Grandson," he said, "I must ask something of you . . . Sing my song," he pleaded with a tired voice. "Sing the song of my people. For I am afraid . . . I can give you nothing in return. Only that I will be with you in spirit."
>
> I could not refuse. I nodded. "I will, Grandfather. I will sing your song."
>
> "You will come to know the ways and powers of my people as your own." He looked long into my eyes . . . there was sadness, wisdom, kindness, and strength, but courage most of all. . . . He turned, without a sound he went into the forest and was gone.
>
> Joseph Marshall III, *On Behalf of the Wolf and the First Peoples*

One major difference between Western and Indigenous ways of understanding the natural world lies in their attitudes toward and response to predatory organisms, such as wolves, bears, big cats, crocodilians, and sharks. In essence, this can be summed up by the observation that despite their tendency to consume large quantities of animal protein, people of the Western philosophical tradition tend to regard themselves as prey. As a consequence, individuals who follow this tradition of fear thus try to exterminate any potential predator. In contrast, followers of Indigenous philosophical traditions tend to regard themselves as predators and show respect for the nonhumans who share their ecological role (Schlesier 1987; Pierotti and Wildcat 1997b).

To anyone who doubts that this characterization is a crucial aspect of Western understanding, including followers of the Western scientific tradition, I refer to recent work by one of the most prominent investigators of the behavior of predatory behavior in mammals, Hans Kruuk. In a recent book that reviewed relationships between mammalian carnivores and humans, Kruuk states, "I will start at the *darkest, most horrifying and negative* side of our relationship with [carnivores], that is their predation on us. They can be *very dangerous enemies*" (Kruuk 2002: 55; emphasis added).

Kruuk's choice of words speaks volumes; to begin with, it is obvious that he considers predators to be enemies of their prey, a concept that flies in the face of all understanding of known ecological relationships. There is no hostility of predators toward prey; it is simply that predatory organisms must take lives to eat. This Western attitude of hostility is used to characterize the relationship between nonhuman predators and their prey. As an example, on U.S. television the reintroduction of wolves into Yellowstone and their interaction with bison has been described as "the rekindling of a blood feud between two ancient enemies" (National Geographic 2006).

When Kruuk discusses the interaction between humans and mammalian carnivores, we are told that, "conservationists often deny that large predatory animals actually kill people, but there is ample evidence that such indignant denial is nonsense" (Kruuk 2002: 55). I know of no one who denies that lions, tigers and other large cats have been known to attack, kill, and consume humans as prey (e.g., Packer 2009), but the evidence is much more equivocal for the carnivorous mammals of North America. Even Kruuk admits that wolves are not known to attack or kill people in North America, despite the fact that there are allegedly many cases of such attacks in Eurasia (75–76). Kruuk never seems to consider the philosophical and cultural basis of this difference between continents, cultural traditions, and human-animal interactions.

One reason for this difference may be that prior to the arrival of Europeans, humans and wolves in North America enjoyed relatively benign relationships between the two species (Schlesier 1987; J. Marshall 1995). In Numic peoples, wolf is considered as a benign creator figure with whom humans enjoyed friendly interactions. There are stories of Comanche hunters finding wolf dens and stopping to play with the pups, while the parent wolves observed from a short distance away (Wallace and Hoebel 1948). Lakota tell a story of wolves encountering a bison killed by humans and sniffing the arrows, looking at the humans, then walking away (J. Marshall 1995: 12). Similar accounts exist for human-lion interactions in Namibia involving the Juwasi people, where a relationship was established that may have existed for tens of thousands of years, whereby these two top predators historically treated one another with respect and deference (Marshall-Thomas 1994).

It is interesting that Kruuk (2002) missed this account by Marshall-Thomas, even though he cites questionable anecdotal accounts of alleged predation. He includes unexplained "disappearances" in Belarus as evidence of predation by wolves (69–70), followed by the statement that he "could not help wondering how much more [wolf predation] would be going on there in the endless wilds of Belarus and Russia" (70). Colleagues who work with conservation and wolf management in Siberia indicate little or no problem with wolves attacking humans (N. Malkov, personal communication), even though Indigenous Altaian herdsmen regularly encounter wolves while tending yak herds in the high mountains to this day (C. Annett, personal communication). This statement of Kruuk's relates to another, more sinister, attitude on the part of Western Europeans and their descendants—the idea that humans from Eastern Europe are barbarians because their forests are more intact, and because, unlike Western Europeans, they have not exterminated their nonhuman predators (Sale 1991; Wolff 1994; Coates 1998). This tradition relates directly to the idea that Eastern Europe is the land of vampires, werewolves, and other half-human "beasts" who prey upon humans (Wolff 1994; Longworth 1997).

In a similar vein, Kruuk seems to view the story of Little Red Riding Hood as an accurate metaphor for wolf-human interactions, stating, "It is perhaps significant that, at least in Europe even these days, the fairy-tale image of wolves is somewhat closer to the truth than the picture painted by some conservationists" (184). In this tradition, Kruuk is not unlike a judge in a case in which I participated as an expert witness. Ruling on whether an individual who owned wolf-dog crosses should be considered as a danger to her community, the judge stated in open court, "There must be something to the story of Little Red Riding Hood." Fantasies of forests filled with dangerous predators longing to feast on human flesh are a long-standing part of the imaginations of Europeans and their American descendants, which leads to both their fear of wilderness and the tendency to regard its inhabitants as bestial.

I suggest that what these attitudes really indicate is the fear-based perception of much of contemporary American and European society concerning large carnivorous mammals. Dr. Kruuk is a resident of the British Isles, an environment from which all carnivores larger than foxes and badgers have been exterminated. A cultural tradition in which humans view themselves as prey is almost certain to create a culture dominated by anxiety and fear with regard to the natural world as described in the previous chapter (Sale 1991; Coates 1998).

The Western perception of humans as prey emerges from the Christian tradition; one key term used in describing Jesus Christ is the "lamb of God." Thus, the "savior of mankind" is regarded as a prey organism to be sacrificed. It is telling that the individual upon whom much of the moral foundation of contemporary Western society is based is perceived as equivalent to a prey organism, which also leads to the view of his disciples and followers as helpless creatures. Christian symbolism is full of shepherd, sheep, and flock metaphors, such as Christ as the Good Shepherd, a Christian congregation as a "flock," and sinners as "lost lambs or sheep." The word *pastor* derives from *pastoral*, which emerges from the metaphor of a shepherd tending his sheep. This metaphor can also be seen in the Hebraic tradition in the willingness of Abraham to sacrifice his son, Isaac. When he decides not to do this he substitutes a lamb. Christians are also told how early adherents to their faith were "fed to the lions" by the Romans.

This attitude does not appear only in ancient documents and stories. Monsters in contemporary films include sharks (*Jaws* and *Open Water* et seq.), lions (*The Ghost and the Darkness*), tigers (*Gladiator*), crocodilians (*Alligator I* and *II; Lake Placid I* and *II; Primeval*), velociraptors and Tyrannosaurs (*Jurassic Park* and sequels). It is telling that in the animated films *The Land Before Time* and *Dinosaur* all of the dinosaurs are anthropomorphized, given humanlike personalities and large eyes, and are capable of speech, except for the predators, who are portrayed as mindless killing machines. In reality, in any ecological community the predators are generally the most intelligent and social creatures, which is why Indigenous people often recognized them as clan members and/or creator figures.

The roster is extended through the use of imaginary beings such as vampires, werewolves, and the Alien. All of these entities are considered to be truly horrible because they are envisioned as predators on humans. The film *Grizzly Man* (Herzog 2005) became a minor hit because it dealt with the consumption of the title character by grizzly bears after fourteen years of foolhardy behavior around these large powerful creatures (Lapinski 2005). The primary outcome of most of these stories and this culture of fear (see also Nicholson 1959; Martin 1999), aside from a fearful populace, has been the extirpation or near extirpation of most large predators from much of their former range in Europe and North America.

As just one example, wolves, including gray wolves and red wolves, *Canis rufus*, used to be distributed throughout most of North America, ranging from Florida to Alaska, and from Newfoundland and Labrador to Texas, New Mexico, and Arizona (McIntyre 1995; Coleman 2004). Today the red wolf is officially extinct in the wild, and the gray wolf in the United States survives only in Alaska, Canada and in small relict populations in Michigan, Wisconsin, Minnesota, Montana, and Idaho (Idaho recently passed legislation designed to re-exterminate wolves in that state). Sporadic attempts at reintroduction in Idaho and Wyoming have met with strong resistance from farmers and ranchers who invariably invoke the image of wolves taking children, even though there are no verified records of wolves in nature ever killing a human being in the history of North America (McIntyre 1995; J. Marshall 1995; Coleman 2004; Mech 1998), even Kruuk (2002) admits to this fact.

Instead, the true relationship between Europeans and wolves is best described as a campaign of unimaginable viciousness directed at wolves by Europeans and their descendants (McIntyre 1995; Coleman 2004). Dates of the exterminations of wolves followed the roots of the Enlightenment (Coates 1998). Wolves were exterminated in England during the sixteenth century and in Scotland by 1684, in Ireland by 1770, Denmark by 1772, Bavaria by 1847, Poland by 1900, France by 1927, and over almost all of the United States by 1950 (Alaska did not become a state until 1959) (Coleman 2004). This extermination effort did not involve simple shooting, as was the case with bison (Isenberg 2000); treatment of wolves by European immigrants was exceptionally vicious and appeared to be motivated by revenge (though for what is unclear), fear, and hatred (Coleman 2004). In 1814 John James Audubon described a Kentucky farmer torturing to death a captured wolf, and farmers and ranchers regularly nailed heads of wolves to public buildings as though they were captured and executed criminals

(Coleman 2004). This behavior continues to this day in the treatment of coyotes, *Canis latrans*, in the American west (author's observation).

This tradition continues in other ways that reveal how little the attitudes of most Euro-Americans have changed since the eighteenth and nineteenth centuries. In the 1960s those paragons of the alternative culture, the Grateful Dead, performed a song titled "Dire Wolf" in which the title character is described as "600 pounds of sin" and the refrain is "Don't murder me, I beg of you, don't murder me." Any shark attack on a human becomes national news and often quickly leads to cries demanding "vengeance" (Casey 2005). The Euro-American mindset seems to want to treat predatory animals as criminals and equate them with psychotic killers. As an example we can look at the series of best-selling novels by John Sandford that refer to murderers or rapists as "predators" and their victims as "prey."

When I was a postdoctoral scientist in Madison, Wisconsin, in 1988, a mentally ill person climbed into the polar bear cage at the local zoo. The polar bear was shot even though it was completely innocent. In Florida a father who dangled his three-year-old over a pool full of alligators and dropped him, tried to have the animals killed. Fortunately the owner of the facility refused. I participated as an expert witness in a civil case where a young father dangled his eighteen-month-old son in front of a wolf-dog and sued for damages when the agitated animal grabbed the youngster by the toe. Grizzly bears and wolves are nearly extinct, but the fearfulness and viciousness of Euro-Americans continues unabated, as illustrated by the 2005 decision by the state of Idaho to kill all wolves in their state and the highly emotional reactions to the film *Grizzly Man*. In the last case, my reaction to this film as a behavioral ecologist was to observe how tolerant the bears were of the photographer and alleged "conservationist" who seemed to be working out his personal demons in front of these large and patient carnivores.

INDIGENOUS AMERICANS AND PREDATORS

In contrast to this litany of sadness and horror, we have the Indigenous American tradition. I have examined hundreds of Indigenous stories that deal with predators, including bears, wolves, coyotes, cougars, eagles, and badgers. In none of these stories are humans the prey of the nonhuman predators, although the opposite is not always the case. The only story I found that alludes to humans as prey for nonhuman animals is a clearly metaphorical creation story from the Northern Cheyenne (Tsistsistas) about a race to decide whether bison should eat humans or humans eat bison. In this story Magpie, as a fellow two-legged, wins the race for the humans, which explains why these people leave portions of their kill for magpies and crows (H. Mann, personal communication; Schlesier 1987).

What is emphasized in Indigenous stories is how much predators are like humans in their behavior and attitudes. The Cheyenne developed their creation stories around wolves who served as their guides and instructors. Animals that hunted were emulated by human hunters (Schlesier 1987: 82; LaFlesche 1995: 132; see Chapter 1 of Marshall 1995 for stories involving Lakota). As part of this tradition Tsistsistas hunters call wolves to their kill or set part of the meat aside for their use (Schlesier 1987). The Blackfeet also have a tradition in which they learned how to hunt bison from wolves (McClintock 1910). In the Shoshone language the name for gray wolf is Numuna (Ramsey 1977: 231), which is virtually identical to the name the Comanche and Shoshone people call themselves (Nuhmuhnuh: Comanche Language and Cultural Preservation Committee 2003). This identification with predators makes Indigenous people feel confident and like they have control over their environment; after all, they are the close relatives and cultural descendants of the most powerful entities in their local ecosystem.

Identification with predators explains why the Indigenous peoples of North America do not have adversarial relationships with wolves, bears, or even cougars (J. Martin 2000; Pierotti and Wildcat 2000). Indigenous peoples function as hunters and recognize themselves as predators within their local ecosystems (McClintock 1910; Tanner 1979; J. Marshall 1995; Pierotti and Wildcat 1997b, 2000; Barsh 2000, n.d.; Henriksen 2009). The Indigenous people of Hawaii similarly view sharks as important spirits and as animal helpers (Kawaharada 1994). In these cases, familiarity breeds not contempt, but respect.

As a consequence of regarding predators as being like humans, almost all Indigenous cultures in the Americas invest hunting with important spiritual and religious traditions (Pierotti 2005a). "The Majority of tribal religions look at religion as a healing and balancing process. Healings are a cooperative enterprise between people, animals, and spirits or powers" (Deloria 1999b: 154). These cooperative or mutualistic relationships between humans and animals underlie the religious and spiritual aspects of hunting for Native American people (Tanner 1979; Schlesier 1987; Harrod 2000; Henriksen 2009). One major component of these relationships is respect, which is a "key operating principle in [Indigenous] religious ideology" (Armitage 1992). "Respect involves maintaining good relationships with the spirit world. . . . Hunters obtain cooperation of the animals they kill by showing respect to their bones and other remains" (Tanner et al. 2002). It is important to understand how relationships with nonhumans became such an integral component of Native American religious thinking and how rituals and traditions that symbolize these relationships continue to function in the religious practices of contemporary Indigenous American peoples.

As Indigenous Americans evolved culturally and ecologically in North America, they learned to obtain food and shelter from the land. To survive and prosper, these peoples had to become effective at hunting, and this continues as a way of life until the present day in some Indigenous cultures (Brody 2000; Nadasdy 2003; Henriksen 2009). Survival, both as individuals and as cultures, depended on their ability to take the lives of other beings. To be an effective hunter required observation of their fellow nonhuman beings, each of which had at least one ability or characteristic that set it apart from other species and enhanced their chances of survival as individuals (Nerburn 1994; J. Marshall 1995; Barsh 2000; Nadasdy 2003). Humans lacked horns, teeth, claws, and the speed and strength of many other species. What humans had was understanding and language, which allowed them to pass knowledge directly from one generation to the next. "American Indians view reality from the perspective of the one species that has the capability to reflect on the meaning of things" (Deloria 1999b: 130). Indigenous Americans survived and prospered by paying careful attention and learning about the strengths and weaknesses of the other organisms in their community, and by developing rituals and traditions related to this knowledge that symbolized the importance of the taking of nonhuman lives. This enabled them to function ecologically as predators rather than as prey; they took nonhumans as food and seemed to avoid being taken by them as food (Martin 1999; Pierotti and Wildcat 2000; Nadasdy 2003; Henriksen 2009).

As knowledge acquired through observation was passed on, and repeated constantly so that the knowledge would be passed on intact, several types of stories emerged that dealt with hunting and the gift of animals (e.g., Schlesier 1987; Harrod 2000). Some stories dealt with animal masters or mistresses, such as White Buffalo Calf Woman. Many dealt with kinship between humans and animals. Others focused on the activity of cultural heroes, and in some cases the gift of animals came from an anonymous or poor individual (Harrod 2000). The boundaries between categories of story were permeable, and different types might be linked in the same narrative.

The notion of kinship created a spiritual conundrum for peoples who depended on hunting as a way of life. The essence of Native attitudes toward other life forms is "kinship relations in which no element of life can go unattached from human society," and manifested itself in "kinship cycles of responsibility that exist between our species and other species" (Deloria 1999b: 131). "Killing animals is a serious business and Kluane people do not engage in it lightly" (Nadasdy 2003: 94). In return, "Every species finds meaning in this larger scheme of things and that is why other species are willing to feed and clothe [humans]" (Deloria 1999b: 149). An Inupiat hunter stated that, "The greatest peril of life lies in the fact that human food consists entirely of souls. All the creatures we have to kill and eat, all those that we strike down and destroy to make clothes for ourselves, have souls, like we have, that do not perish with the body, and which must therefore be propitiated lest they should revenge themselves on us for taking their bodies" (Ivaluardjuk, cited in Rasmussen 1929).

If nonhumans were understood to have "characteristics similar or equivalent to those of humans, how were humans to understand what it meant to kill animals and consume their flesh?" (Harrod 2000: 46). Along with the unpredictability of environmental conditions, this dilemma is one of the defining elements of Native American religious thought. Many rituals and traditions stem from practices developed to provide an ethically satisfying resolution to the taking of other lives because participation in rituals and ceremonies builds understanding (Cajete 2000). It is not widely recognized, but many contemporary religious practices of Indian people stem from rituals originally developed for hunting, for example, pipe ceremonies and the sweat lodge transformed into practices that address the needs of contemporary Indians (Harrod 2000).

Indigenous American peoples were fully aware that they were not the only organisms who hunted as a way of life. They observed that other organisms killed and ate the plant eaters. Wolf, cougar, bear, badger, and eagle were fellow hunters from whom much could be learned. These animals took on sacred significance—for example, they represent the guardian spirits of five of the six regions (directions) in Zuni religious thought: wolf (east), badger (south), bear (west), cougar (north), and eagle (sky) (Cushing 1988; the sixth guardian, mole, is also a predator, but not a large carnivore). Predators were recognized for their power, and humans recognized a kinship with them since humans also took animal lives for food. "From the dawn of our spiritual and psychological being our closest relative in the wild has been Makuyi—In English, Wolf" (Jack Gladstone, Blackfeet, cited in McIntyre 1995). Stories exist from the Coast Salish of humans who hunt with wolves and share killed deer with the wolves (Barsh n.d.).

One important feature of including nonhumans as community members is that it allowed Indigenous Americans to identify with and respect predators, since they knew how difficult it is to take the lives of other individuals (Marshall-Thomas 1994; Tanner 1979; Marshall 1995; Harrod 2000; Henriksen 2009). In this intellectual and spiritual tradition it is recognized that predation is not an activity that involves hostile intent, and that predators may feel strongly connected to the prey after they have taken its life. For example, cougars (*Felis concolor*) and leopards (*Panthera pardalis*) groom their prey after a kill as if the prey were a young member of their own species. This last point is interesting; group-foraging carnivores, like wolves and lions, do not seem to have rituals in feeding, whereas solitary carnivores, like most other big cats, seem to have gentle appearing rituals associated with taking prey between killing and eating. Cougars, in particular, seem to be very precise and careful in how they feed and how they treat their prey (Baron 2004). For example, " . . . a large male puma . . . just killed a male bighorn sheep (that was) lying dead on its left side. The puma lies down on its right side, face to face with the sheep, gazes into its eyes for a moment, then reaches out his paw and tenderly pats the sheep's face as a kitten might pat its mother" (Marshall-Thomas 1994: 25).

Western science seems to be rediscovering principles that are part of the TEK tradition with regard to predation. Until the 1930s the U.S. government funded the Division of Predator and Rodent Control, also known as "gopher chokers" and the "poison brigade" (Dunlap 1988), which resulted in widespread efforts to eradicate the wolves, coyotes, cougars, and bears in the United States. This attitude did not begin to change until the 1930s when Aldo Leopold's graduate student Paul Errington established that predators did not have a harmful impact on populations of game birds and mammals (Errington 1934, 1937, 1967). Although Errington was a dedicated hunter of European heritage, he also understood predators as part of the system. He learned to respect coyotes as a boy running a trapline in South Dakota, and he understood that many parts of the American West were damaged by Euro-American development (Errington 1973). Errington's results, combined with the spectacular failure of deer management strategies after the extermination of wolves and cougars on the Kaibab Plateau, helped to slow the drive toward extermination of predators that was driven by ranchers and their allies in the U.S. Congress (Dunlap 1988).

In recent years the role of predators in natural systems has received greater respect and understanding. It has now been established that "real food webs are structured such that top predators act as couplers of distinct energy channels that differ in both productivity and turnover rate" (Rooney et al. 2006). High biodiversity is associated with stability and health, and diverse ecosystems are those that have their predator populations intact. Indigenous people have known the truth of this for a long time. In 1988 I interviewed for a position as a biodiversity expert with the U.S. Environmental Protection Agency. During the interview process I was asked if I could name one criterion that would show that biodiversity in a system was healthy and intact. Following the example of Paul Errington, I replied, "Certainly. If a community has a healthy population of its top predators it will always be stable and diverse." This answer clearly did not please the Euro-Americans who were interviewing me. They pressed me to come up with alternative criteria. I insisted that this was the primary criterion and that it was relatively easy to assess, but it would be hard to achieve this in a society where top predators were hated and feared. It is rewarding 20 years later to see that theoretical and empirical studies support the point I made based on Indigenous principles and TEK (Holt 2006).

RELATIONS WITH SPECIFIC PREDATORS

Wolves

One particular predator was of great cultural and spiritual significance to many Indigenous American peoples, such as Comanche, Shoshone, Blackfeet, Cheyenne, Lakota, and Coast Salish (Schlesier 1987; J. Marshall 1995; Barsh 1997, n.d.). This was Wolf, who was found throughout North America, lived in family groups, and was not strong or swift enough to kill large prey alone. Wolves working cooperatively as a group, however, could bring down even large plant eaters, and thus served as models for the concepts of community existing at both the single species and ecosystem level (Bruchac 2003: 159). Like humans, wolves associated with and maintained cordial relations with other species, such as ravens (Barsh 2000; Heinrich 2000). Western scientists studying wolves have realized that Indigenous Americans have far greater knowledge of the behavior and ecology of wolves than Western science and have turned to Native people to help them in their study of these animals. For example, an Alaskan wolf biologist has described the difficulty he had in locating active dens until he turned to local Inupiaq hunters for help (Stephenson 1982).

Understanding wolves and their relationships to other species solved a behavioral conundrum in Western science. The superb field biologist Bernd Heinrich studied Common Ravens, *Corvus corax*, feeding on carrion in the forests of New England (Heinrich 1989). Heinrich was curious that in his study populations in Maine, ravens were shy near carcasses; in fact, they were so shy that he feared they "might be almost paralytically afraid of dangerous ground predators" (Heinrich 2000: 231). The idea that ravens feared mammalian predators was suggested by a prominent behavioral ecologist (Heinrich 2000: 217). To test this idea, Heinrich traveled to Nova Scotia to observe a captive wolf pack and their interaction with ravens. He put down two piles of meat at the same time, knowing that based on their traditions, the wolves would feed as a group on only one of these piles. Local ravens thus had a choice, to feed with the wolves or to feed away from them. In every case the ravens chose not to avoid the predators, but to feed with them.

Subsequently, Heinrich traveled to Yellowstone National Park after the reintroduction of wolves in 1994 and observed numerous carcasses, both attended by wolves and unattended. In all cases the carcasses attended by wolves also had ravens, magpies, and eagles feeding from the carcass. Not once did the ravens hesitate to go in and feed from a carcass or meat pile attended by wolves (Heinrich 2000: 232–34). In fact, the ravens hung around the wolves even when there was no food present. This suggests that, rather than fearing predators, it is the absence of wolves that makes ravens shy and much more nervous around the body of an elk or deer. When wolves are present, ravens and other birds are more confident and relaxed (Heinrich 2000). The fear and shyness of ravens in Maine resulted not from possible threats from terrestrial predators, but from the absence of their fellow predators. For ravens there should not be carcasses around without their friends the wolves. In this context the Tsistsistas tradition of "calling" ravens, coyotes, and foxes to share in their kills (Schlesier 1987: 82) can be recognized as part of a tradition that may go back millions of years.

Another misunderstood relationship comes from the TEK-based idea that badger and coyote were "friends" and hunted together. Western ecology, driven by the idea that competition among species drives community dynamics, categorized the relationship between coyote and badger as competition between two predators (Minta et al. 1992; film footage of this relationship can be seen in *Yellowstone: Realm of the Coyote* [National Geographic 1995]). In contrast, the assumed competition is emphasized in a documentary shown on *Animal Planet (Badgers: Dishing the Dirt*, April 23, 2005) that includes faked footage to imply that badgers and coyotes are rivals. Empirical study by Euro-American scientists, however, revealed the empirical basis of the story that they are actually friends. Coyotes and badgers spend a lot of time wandering around together, but when they see a ground squirrel, coyote gives chase. If the squirrel goes into a burrow, badger will dig up the burrow, or both will dig together. If the squirrel stays in the burrow, badger will often get it and have a meal. If the squirrel attempts to escape by using another burrow exit, coyote often gets it and has a meal. Both coyote and badger catch more squirrels when they hunt together than when they hunt alone (Minta et al. 1992).

Given this ancient heritage, the reliance of Plains tribes, such as the Cheyenne, Lakota, Crow, and Comanche, on wolves as models is not surprising. The Innu of Labrador have stories that show how wolves hunt in a way that shows respect, whereas others do not (Henriksen 2009: 96–99). The weapons of wolves were "formidable, but the first people saw that they were of little use without endurance, patience, and perseverance . . . qualities the first peoples could develop in themselves" (Marshall 1995). Several Indigenous peoples felt their connection to wolves was strong because wolves even instructed them in methods of hunting (Schlesier 1987; Barsh n.d.). Wolves were also respected for their alertness, endurance, and ability to be part of close-knit group, but also for their ability to function well when they were alone. Leaders of war parties were "spoken of as wolves because they are men of great fortitude . . . who, like wolves, are ever alert, active, and tireless . . . who can resist the pangs of hunger and the craving for sleep" (LaFlesche 1995: 132). Most important, however, was that if people were to emulate the wolf, they also had to exist, like the wolf, to serve their own social community and the local ecological community. "Understanding this reality made them truly of the earth, because every life ultimately gives itself back to the earth" (Marshall 1995). A prominent Western behavioral scientist has suggested that human social evolution may have been strongly influenced by observation of, and even cooperative hunting with, wolves over the last several thousand years (Schleidt 1998). Another investigator argues that the reason modern humans supplanted Neanderthals was because of the presence of wolves/dogs as early warning systems and cooperative hunting partners, which Neanderthals did not have (Coren 2008).

Bears

Bears are also of symbolic and spiritual importance to numerous peoples among Indigenous Americans (Rockwell 1991). As with wolves, bears in some cultures were seen as almost interchangeable with humans. Bears resemble humans, especially after they have been skinned, and bears are known to regularly stand and even walk upright for short distances. The Blackfeet word *o-kits'iks* refers to both the human hand and the paw of a bear (Grinnell 1962). The Tlingit people of southeastern Alaska refer to grizzly bears as half human. As with Numic peoples and their word *Numuna*, the Ojibwe word for bear is *anijinaabe*, which is virtually identical to their word for themselves, *Anishinaabe* or *Anishinaabeg* (the latter is the preferred spelling of author Louise Erdrich). This metaphor is given flesh in Erdrich's novel *Tracks* (1988) and Gerald Vizenor's novel *Dead Voices* (1992), where the main character is an Anishinaabeg woman who may or may not be a bear.

To Indigenous peoples of the northern American forests, bears are sometimes regarded as creator figures, partly because the highly protective nature of female bears makes them seem exemplary mothers. In these cultures diets of both humans and bears overlapped extensively. These two species often foraged together in the same areas, and humans learned which plants were edible from close observation of the foraging choices and tactics of the bears. Thus, in many tribal traditions the bear is recognized as the "plant gatherer," bestower of the secrets and mysteries of plants (Rockwell 1991: 6). In the Hidatsa and the Blackfeet there is also a tradition that bears taught people how to trap eagles (McClintock 1910; G. Wilson 1929).

In the Pacific Northwest, bears are considered to be close relatives of the Bear clan of the Tlingit people. At some levels this appears to be more than metaphor. A story is told of a Tlingit elder from the Bear clan who died. As his body was being carried by truck to the village of Angoon, "a half dozen or so bears materialized out of the woods and lined the road from the dock to the village . . . As the truck [carrying the body] rolled by, some [bears] actually stood up" (C. Martin 1999: 37). Although to a Western trained scientist this story may be hard to accept, I had this story confirmed to me by a student at Northwest Indian College, who said that the elder involved had been his relative.

Such relationships, built on respect built up over long periods of time, are not part of the Western scientific experience, yet even Western scientists are sometimes privileged to observe such phenomena. In the Kalahari Desert Indigenous people apparently had a mutually respectful relationship with lions, such that lions would honor kills made by humans and even get up and walk away when the humans came to claim kills they had made (Marshall-Thomas 1994). Joseph Marshall tells a similar story with regard to a female wolf who sniffed the arrows in a dead bison, looked at the hidden hunter, and then walked away (Marshall 1995).

To Indigenous people, the interesting thing about the story told in the movie *Grizzly Man* (Herzog 2005) was not its gruesome ending, but the fact that a Euro-American who often behaved in a disrespectful manner around bears was able to survive for fourteen summers around the largest land carnivores on earth without coming to harm until he got caught in a territorial struggle between an adult and sub-adult male bear. It is obvious from the film that the bears showed remarkable tolerance of Treadwell and his often bizarre behavior.

In contrast to bear and wolf, cougar seems to feature prominently in stories of Indigenous Americans only in the Southwest where cougar is a common predator of deer, just like humans. It is probably harder for humans to identify with this secretive and solitary hunter. Cougars, at least in their rare melanistic form, also feature in the Choctaw tradition of *nalusichito*, the soul eater (Owens 1992b, 1998), but this is not a positive image of a creature to be emulated. The main place where cougar features prominently is in functioning as both the spirit of the hunt and the guardian of the North in the tradition of the Zuni and other Pueblo peoples (Silko 1977). Cougar (or puma) were also symbols of courage in Osage ceremonial practices (LaFlesche 1995: 95). Cougars were recognized and respected for their power as predators, "[the cougar] never makes noise because he has the right sort of moccasins" (Eastman 1904: 242). Because of their solitary nature, however, cougars were not a species readily emulated by the group hunting and foraging humans, especially on the open plains where cooperative hunting is important.

Another aspect of Indigenous American philosophy that is related to the societies of socially hunting predators was an emphasis on the importance of multigenerational cultural traditions. In these traditions, both elders and children are seen as sacred (Bruchac 2003: 143). It was recognized that elders and offspring were likewise important in animal societies.

In contrast, the Western scientific tradition of behavioral ecology seems not only to misunderstand the social interaction between different species but also to ignore the importance of elders and offspring in the hunting tradition. For example, a conceptual model of cooperative hunting in carnivores assumes that older animals and younger animals are "cheating" the present generation of hunters by feeding from a kill when they did not directly participate in making the kill (Packer 1988; Packer and Ruttan 1988). These investigators ignore the fact that to the present generation of hunters of any cooperatively hunting species, be they human, wolf, or lion, older and younger individuals represent the previous and the next generation of hunters. Today's hunters were youngsters when the elders were the hunters, and they will be the elders when the young ones of today become the hunters. As an example, see the relationship described in Indigenous terms by a Lakota elder:

> Your way harms children because it confuses being useful with being important . . . children are important because of where they stand in the circle of life. This is something we have tried to share with your people. We have tried to remind you that life is not a straight line from birth to death, but a circle where the young and old hold hands . . . (Nerburn 2009: 303).

The only difference between the Lakota and the lions is that the lion elders and young ones feed from the same kill, rather than metaphorically holding hands, but the point made is important in that the Western-derived model (metaphor) of the behavioral ecologists ignores the cyclical multigenerational nature of cooperative hunting societies and is based on a naïve metaphor derived from contemporary Western socioeconomic models of competitive individuals, rather than real functional social systems.

These cooperative, even "friendly" relationships between species and across generations are literal illustrations of the important principle shared by most Native American religious or spiritual traditions, that *all things are connected*. This is a spiritual acknowledgment of the realization that no single organism can exist without the connections it shares with many other organisms. Taking lives of other organisms and consuming their tissues in order to sustain one's own body tissues establishes connectedness. By eating parts of other organisms, you demonstrate empirically that they are made of the same materials of which you are made (Pierotti and Wildcat 1997b, 2000, 2001). Christianity employs a similar principle in its communion rituals as a way of establishing links between their "savior" and contemporary humans.

PHILOSOPHICAL AND ECOLOGICAL PRINCIPLES

Recognizing connectedness did not mean, however, that animals or plants should not be taken or used for food or clothing. Instead, each taking was accompanied by recognition of the fact that the take represents loss of life to a fellow being whose life had meaning on its own terms (Taylor 1986, 1992). This recognition led to two ethical and spiritual conclusions based on the concept of respect: (1) lives of other organisms should not be taken frivolously; and (2) other life forms exist on their own terms and were not put here only for human use (Pierotti and Wildcat 2000, 2001; Henriksen 2009). Indigenous Americans recognize that the lives of animals and plants exist on their own terms and have value independent of any we human beings place on them. Despite this value, being taken as food is a common fate of species within their natural environment, such as deer, salmon, and bison. Consuming the flesh of these animals establishes the connectedness that is such an important aspect of spirituality.

The lives of human hunters and their families depend upon taking the life of the animal. By giving up its life an animal was making a profound sacrifice, which required thanks. In such situations, Indigenous Americans truly understood themselves as predators, part of the world of the prey and connected to the prey in an experiential sense, and recognized that a predator can feel most connected to the prey when taking its life. As an example, the Kluane people of the Yukon "see themselves as enmeshed in a web of reciprocal social relations with animal people. They see success in hunting both as evidence that they have properly maintained these relations and as an indication of their obligation to continue doing so. To respect animals is to conduct these relations properly and to live up to one's social obligations to these important other-than-human people" (Nadasdy 2003: 94).

Some scholars do not seem to understand the nature of these obligations. It has been pointed out that modes of relationship within an animistic worldview take on two main forms: reciprocity and predation. Some argue that reciprocity "implies the *need for a total balance* in the exchange of substance between human and nonhuman beings," whereas predation presumably involves "*aggressive exploitation* of nonhuman beings as are useful" (Harkin 2007: 217; emphasis added). Harkin goes on to argue that humans "never repay their debts to animals and thus never achieve genuine reciprocity . . . humans appear almost exclusively as predators rather than as equal partners" (218). This line of argument seems wrong at least two levels. First, Harkin seems to completely ignore the idea that ceremonies and rituals designed to placate the animals taken are as close to full reciprocity as any predator can get. It seems as though Harkin expects that humans should offer themselves to prey species as sacrifices to achieve true reciprocity. Following Harkin's logic, no predator exists in a co-evolutionary or egalitarian relationship with prey species, which, although it ignores evolutionary ecology (see Chapter 4) is a position regularly occupied by scholars who argue that Indigenous people do not really understand how to maintain a sustainable relationship with nature (see also Krech 1999a; Kay 2002a). More importantly, it shows that Harkin does not seem to understand the nature of reciprocity in natural systems. One of the top students of reciprocity in nonhumans has pointed out that "parental care doesn't need to favor one's own offspring (and) assistance given to others doesn't require the actor to know if, when, and how they will benefit from it" (de Waal 2009).

Even Harkin's understanding of the situation among First Nations of the Pacific Northwest seems naïve. In direct contrast to Harkin is the situation described by Langdon (2007), who argues that evidence from the Gulf of Alaska indicates that, " . . . over a period of eight hundred years, the Karluk River and lake system on Kodiak Island supported four to six million sockeye salmon annually, well above levels present over the previous twelve hundred years. This occurred when . . . human (Indigenous) populations were at their prehistoric maximum. It is possible that patterns of human engagement . . . enhanced or optimized salmon abundance through conscious intervention" (269). In such a situation human enhancement would be a clear example of reciprocity that was directly linked with predation. In a similar fashion, regular and systematic burning of habitats by Indigenous peoples creates better habitats for deer, moose, elk, bison, and many food plants, thus establishing a benefit that hunters perform for possible food species.

This connectedness that Indigenous American peoples feel toward their prey leads to different perspectives on several key issues with regard to fisheries biology. For example, Native people find that many fishing regulations make little sense in terms of the way they live their lives, such as the situation regarding Anishinaabe and walleye in Wisconsin (Nesper 2002). Fishing regulations are designed to regulate greed on the part of fishermen, yet Indian people feel that the respect they hold for the fish means that they would not overexploit the resource; thus they may also be frustrated by regulations designed to solve problems for which they feel no responsibility (Pierotti and Wildcat 1999a).

On a similar theme, catch-and-release fishing, which is considered a major tool for conservation by non-Indigenous sport fishermen, may be regarded as "playing with the fish" because it shows no respect for the fish and the importance of its life (Nadasdy 2003: 81–83). Indigenous people feel that if a fish is caught, that fish should be eaten; otherwise any suffering experienced by the fish during its capture is for no purpose (Pierotti and Wildcat 1999a). In showing respect for the sacrifice made by fish, Indigenous people reveal the true basis of their reciprocity, which, despite the arguments posed by the likes of Harkin (2007), are based on an specific ethical system that derives from Indigenous experience and not upon the exploitative ideas that come from the Western conservationist tradition (Brightman 1993; Krech 1999a; Kay 2002b). As a final example, hatcheries, which are an important management tool in Western conservation practice, are often considered to be the equivalent of farms; but they show no respect for the fish as individual beings and turn out individuals that may not be able to function as true fish once released into nature. "Hatchery reared fish don't taste right" (B. Frank, personal communication).

It is possible to take the life of another creature while respecting and admiring that creature and while recognizing the value of its life by taking only one life at a time (Pierotti and Wildcat 1999a; Pierotti 2005a). This appears to be an important and emotional experience for respectful hunters and fishermen, regardless of ethnic heritage. If animals are raised in unnatural circumstances, as in feedlots or hatcheries, and their lives are taken en masse, as in slaughterhouses or fishing trawls, it becomes much more difficult to show respect for these lives. Each individual becomes submerged within the mass. To ignore this scale of suffering, non-Natives convince themselves that these other creatures lack feeling, including lacking the ability to experience pain. Thus they come to regard the creatures' lives as having no value beyond a monetary one.

In contrast, one tradition that illustrates the importance of fish as prey to Indigenous Americans can be seen in the First Salmon ceremonies of tribes in the Pacific Northwest and southeastern Alaska. When a salmon run began, no one was allowed to fish until the first catch was ceremonially welcomed. The fish was bathed, painted with red ochre, and sprinkled with bird down. The fish was ceremonially carried to the place where it would be prepared, laid on a bed of ferns, and dabbed with ochre. Women butchered the salmon, carefully removing the flesh from the spine in one piece so it could be roasted over a firepit. Prayers were offered so that the fish would look kindly on the people and return in great numbers (Amoss 1987; McKervill 1992; Berman 2000). All of the flesh was eaten, sometimes by the entire community, sometimes only by elders and children (cf. the discussion of Western theories about group hunting above). The bones, head, and entrails were carefully collected in a basket or mat and reverentially deposited in the river or the sea where the fish had been caught. This was intended to convince the salmon that it should come to life again and lead others of its kind to the fishing sites (Raban 1999: 403).

Anthropologists have interpreted this ceremonial treatment of salmon as a way of recognizing the spiritual kinship between the people and the nonhumans who shared their world. Although this interpretation is fine, it proves to be overly simplistic, because it assumed that this ceremony simply acknowledged the consistently predictable and reliable natural abundance of the fish. As an example, Harkin (2007) uses the First Salmon Ceremony as an example of how humans cannot reciprocate, and his interpretation is based on the Western-derived "Balance of Nature" metaphors based on an assumption of stable ecological communities. This can be seen in Harkin's statement that "Salmon were thought to come from a place of infinite wealth and to represent a steady supply of food, provided, of course, that the rituals were completed and the bones returned to water" (Harkin 2007: 228).

Recent studies that analyze bottom sediment from coastal waters of British Columbia and southeast Alaska have shown that far from being consistently predictable, there has been enormous variation in the availability of salmon over the last two millennia (Raban 1999: 400). Year-to-year fluctuations in salmon runs typically show relatively little variation, but over 20 to 30-year periods the availability of salmon approximates a wavelike pattern during which salmon populations peaked, declined, remained in a trough for several years, came back, and then diminished again after crests that lasted five to ten years (Raban 1999). In the early sixteenth century there were big runs, followed by a steady fall to the middle of that century. In the early eighteenth century populations were low, but they recovered near the end of that century so that when Europeans arrived there appeared to be a time of abundance for salmon, which the Europeans apparently assumed was the normal state. There was another peak in the late nineteenth century that seems to coincide with the beginning of commercial salmon fishing. Starting in the 1950s until the present there has been a consistent low in salmon numbers, which corresponds with a decline in the salmon fishery (Raban 1999).

This pattern emphasizes two points. First it shows that the commercial fishing industry needs to be aware of the cyclic nature of the environment. Western economic models assume relatively constant conditions, which is why they set maximum sustainable yield (MSY) at approximately 50 percent of the theoretical carrying capacity. Because Euro-Americans assume the environment is stable, they believe that carrying capacity is a real and constant value, rather than fluctuating and unpredictable (Wilson 2005). In the real world there is no constant *K* and environmental conditions change regularly in unpredictable fashion. Thus, it is probably not a good idea to plan on salmon fishing as a lifelong career because the variable nature of the fish populations means that it is fated to be a boom and bust industry. Second, this illustrates once again how Indigenous ceremonial and spiritual practices acknowledge the changeability of the environment and the real possibility that key food resources might decline or even disappear at times. In each village during peaks, "there would have been elders who could remember when so few fish came to the traps . . . and in times of dearth there would be old men prattling about the great salmon runs of their youth. . . . Fish were honored because there was a real and proven likelihood that they might not come back again in sufficient numbers to feed the tribe. . . . The whole corpus of Northwest Indian literature adds up to an epic parable about the capricious and untrustworthy nature of nature" (Raban 1999: 404).

Again we see an Indigenous culture that bases its primary ceremonies and rituals on recognition of the unpredictability of the natural world. Food plants and animals had to be treated with proper respect, for if they were not they could show their displeasure by leaving, plunging the human population into conditions of need and, in extreme cases, even of starvation. Harkin (2007) acknowledges that starvation occurs reasonably often in Northwest coastal peoples, but as indicated above, he assumes that these peoples were incapable of recognizing the relationship between shortages of food and variability in salmon runs. This type of attitude has recently manifested itself in the case of spurious and misleading attacks on Indigenous knowledge of ecological conditions and their variable nature (e.g., Krech 1999a; Kay 2002a, 2002b; Harkin and Lewis 2007).

Perhaps the best way to think of this knowledge borne of experience is that Indigenous people *lived their lives as though the lives of other organisms mattered*. They experienced other creatures in their roles as parents, as offspring, and ultimately as persons within a shared community, and they realized that their own lives were intimately intertwined with those of these other organisms. One indication of this mutuality was that it was always possible that the nonhumans upon whom the human culture depended could go away. Realizing this provides an explanation for First Salmon ceremonies, the prayers and thanks to deer, moose, caribou, and bison, and the treatment of bear and wolf as honored teachers who helped humans figure out how to feed and care for themselves. From this perspective, human beings are not the measure of all things but exist as but one small part of a very complex ecosystem, in contrast to the Western view that places human beings above the rest of nature.

In the tradition of Indigenous Americans, thinking of yourself as a predator makes you less fearful and more confident. The necessity of placating nature keeps people respectful, even if in some ways the relationship between predators and prey represents a fine balance between relying heavily on specific populations of nonhumans while also adjusting to their tendency to vary dramatically in numbers.

It appears, however, that if disease wipes out your community, thus decimating its traditions that such incomprehensible loss creates the opposite state of mind (see Chapter 2). As a consequence of the widespread epidemics that swept North America after the arrival of Europeans, Indigenous people came to think of themselves as victims, especially of the Europeans and their religious and philosophical traditions and they may have become involved in patterns of exploitation that led to local extirpations of some nonhuman populations (Martin 1978; Feit 2007). Another odd consequence of this change of perspective is that today militants who try to represent the Indigenous mind-set, such as the Lakota poet-musician John Trudell and political and social commentator Ward Churchill, now refer to Euro-Americans as "Predator" (Churchill 1995), which to me sadly indicates that, like Euro-Americans, many Indigenous people have now come to think of themselves as "prey."

4 Metaphors and Models

Indigenous Knowledge and Evolutionary Ecology

All Nature is perverse and will not do as I wish.

Charles Darwin, *Letters*

We are beginning to believe more and more that we are some sort of creation of our own mind, our own intellect, of the reasonable aspect of humanity, of logic. We are becoming aliens on our own planet.

Farley Mowat (1997)

As I have argued in previous chapters, Indigenous peoples in North America have relatively sophisticated concepts with regard to the functioning of ecological systems, and these concepts also differ in some important ways from those of Europeans (Minta et al. 1992; Anderson 1996; Barsh 1997, 2000; Huntington and Myrmin 1996; Pierotti and Wildcat 1999b, 2000). These concepts were often presented in a metaphoric fashion, which has allowed some Europeans to try to dismiss them as mere "stories" or "legends" (Deloria 1992; Pierotti and Wildcat 2000).

It must be recognized, however, that metaphor is also an important aspect of the Western scientific tradition. Metaphors dominate both biological research and the way biological phenomena are understood under the Western tradition. In *The Structure of Scientific Inference*, Mary Hesse states: "Contemporary philosophers have argued that scientific understanding, like all human understanding, proceeds by way of providing metaphorical redescriptions of phenomena" (Hesse 1974: 62; see also Lewontin 2001). Given this recognition of the importance of metaphor in the Western tradition, the power of TEK as a way of knowing therefore cannot be dismissed simply because knowledge is sometimes coded in metaphor.

The use of metaphor can be illustrated through the importance of predators— wolf, coyote, bear, badger, cougar, raven—in a wide range of stories (e.g., Mourning Dove 1990; Rockwell 1991). Indigenous tradition recognized humans as predators rather than as prey and did not view other predators as rivals but as powerful entities with whom they shared an ecological role (Pierotti and Wildcat 2000).

These metaphors originate from the Indigenous philosophical and spiritual tradition of deriving knowledge from understanding how ecosystems and populations functioned based on careful observation of relationships, which led to the principles that *All things are connected*, and *All things are related*. The Lakota version of the latter principle, *Mitakuye oyas'in* (literally "all my relations"), is the traditional thanks given at the close of prayers or religious ceremonies. "All my relations. That means everything in the world—the plants, the animals, the sky, the trees, the rocks—everything. When you feel that everything is your relation, you feel that everything is connected" (Lakota elder cited in Nerburn 2009: 302). Such statements demonstrate how spiritual tradition can be used to reinforce a system of knowledge, while also illustrating the relationship between ecology (connection) and evolution (relatedness) as well as any Western text (Pierotti and Wildcat 1997a, 1997b; Marshall 2001). All Indigenous peoples of whom I am aware consider humans and nonhumans to be related, which is also a principle that lies at the foundation of evolutionary biology (Mayr 1997; Pierotti and Wildcat 1997a, 1999a, 2001).

EVOLUTIONARY ECOLOGY: WESTERN SCIENCE AND ENVIRONMENTAL VARIATION

One of the most active and insightful fields of study in Western science over the last two or three decades has been the discipline called evolutionary ecology, which involves the application of ecological principles to understand evolutionary phenomena as responses to specific sets of environmental conditions. It also explores how changes in population responses can be interpreted in response to variable environmental conditions (Stearns 1992; Bulmer 1994; Weiner 1995; Hoffman and Parsons 1997; Houston and McNamara 1999).

Like much of science, evolutionary ecology depends heavily on theoretical and mathematical models, including models for population dynamics, life history evolution, foraging theory, natural selection, and the evolution of mating systems and social behavior (Bulmer 1994; May and McLean 2007). Some investigators in this field have even argued that the theoretical and empirical aspects of evolutionary ecology should be considered to be equivalent (Bulmer 1994). As I have argued, these models should also be considered as metaphors or analogs for the actual dynamics of populations that consist of individuals, and for population-level responses to environmental variation. As one example, Game Theory has also been employed to model behavioral strategies in free-living animals, an approach that has showed some interesting results in revealing differences among individuals (Pierotti and Annett 1994). In most cases, however, evolutionary ecology is based upon the perspective that the best way to understand evolution and population dynamics is to understand how the average or "typical" individual functions (e.g., Krebs and Davies 1984).

Since the Enlightenment the traditional model of nonhuman organisms has been the Cartesian metaphor of the machine, which considers organisms to be made up of parts. This metaphor emerged from the Materialist worldview and the links between economics and science that developed in Western Europe during the sixteenth and seventeenth centuries (Worster 1992). Following Cartesian thinking, the most effective way to understand the organism is by understanding its constituent parts and how they fit together (Coates 1998; Lewontin 2001). Discussing Western science, Jacques Monod contends: "The entire system is totally, intensely conservative, locked into itself, utterly impervious to any 'hints' from the outside world. . . . This system obviously defies 'dialectical' description. It is not Hegelian at all, but thoroughly Cartesian: The cell is indeed a machine" (Monod 1979: 108). The Cartesian approach creates a variety of conceptual problems in biology. For example:

> The task of evolutionary biology consists in providing a material, historically based explanation for current or extinct biological forms . . . this task often involves the decomposition of organisms (or ecosystems for that matter) into simpler constituent parts, which can then be analyzed in greater detail. This approach, broadly known as *reductionism*, has proven enormously successful. Incautiously used, however, reductionism has led to a dangerously oversimplified shorthand in evolutionary biology—an interpretation in which organisms are decomposed into features and features are accounted for by single evolutionary causes . . . a bird becomes no more than a set of individual features in search of explanation. . . . While these explanations capture something crucial about the evolution of birds, they deliberately avoid what all biologists know to be true: Organisms are more than collections of independent features and features are rarely shaped by single causes. (Dorit 2007: 234)

A similar reductionist approach has often been taken by Western science to explore functioning of ecosystems, for example, emphasis on trophic levels, producers, and consumers. Under such models, the investigator must take the system or organism apart to study how it functions. Reductionism has led to the emphasis in Western science on dissection and partitioning. The manner in which the machine model is employed in evolutionary ecology is through examination of the functioning of specific adaptations, which are equivalent to pieces, either of behavior or anatomy, for example, choice of diet was originally considered to be independent of mate choice or habitat choice (Stephens and Krebs 1986; but see Pierotti and Annett 1987, 1991, Annett and Pierotti 1999 for alternative concepts).

Combined with the machine metaphor, the use of averages as a stand-in for individuals leads most investigators in the Western scientific tradition to assume that any species under study can be represented by a typical individual, which is constructed of a combined set of adaptations, each of which solves a problem. This point of view is clearly illustrated in the following quote from the eminent student of insect behavior and author on biodiversity Edward O. Wilson: "In order to use models of population genetics . . . I conjectured that there might be single, still unidentified genes affecting aggression, altruism, and other behaviors . . . such traits are usually controlled by multiple genes, often scattered across many chromosomes, and that *environment plays a major role in creating variation among individuals* and societies. Yet, whatever the exact nature of the genetic controls, I contended, the *important point is that heredity interacts with environment to create a gravitational pull towards a fixed mean*" (E. Wilson 1994: 333; emphasis added).

Although Wilson acknowledges the role of the environment in creating individual variation, he still makes two key assumptions. First, he assumes that aggression, altruism, and so forth are distinct traits (parts) that can be pulled apart and analyzed separately, rather than choices from among a spectrum of behaviors that can be used to fit specific social dynamics. In fact, he conjectures that these may be the result of single genes. Second, he assumes that the interaction between genes (heredity) and environment leads to a fixed mean—an average individual who is favored by selection. One consequence of this type of thinking is that in evolutionary ecology separate analyses are typically made of mate choice, reproduction, foraging and diet, sex ratios, and aggressive behavior, as if these entities were all separate from and completely independent of one another (Krebs and Davies 1984; Bulmer 1994).

Over the last three hundred years this reductionist machine-based analytic approach has been successful in explaining some aspects of nature, but has also led to an oversimplified view of the relation of parts to wholes and causes to effects (Lewontin 2001; Dorit 2007). Emphasis on simplified systems for study results because scientists are opportunistic and tend to pursue only those problems that are most likely to yield to their methods, which leads them to ignore more complex and intractable problems, like social or ecological dynamics. As Thoreau phrased it, they are looking past nature, rather than directly at nature. The Western scientific tradition assumes that more difficult questions will eventually be solved through use of the same basic methods that led to solutions of simpler problems, even though the questions involved may be qualitatively, rather than quantitatively, different in nature (Lewontin 2001).

As an example, if knowledge of DNA sequence leads the investigator to understand how information on protein structure is stored, it is assumed that knowledge of molecular structure will eventually allow us to understand how the brain functions because the brain is simply an aggregation of proteins and nucleic acids. This approach ignores the fact that not all information about protein structure is actually contained in DNA—only the sequence of amino acids is revealed by understanding the sequence of DNA codons (Lewontin 2001; Ball 2001). In reality, three-dimensional structures of proteins result from the folding of polypeptide chains and at times the same chain of amino acids may produce proteins with different shapes and functions. The manner in which these chains fold is driven by the shape of molecules and also by bonds between the molecules that are not readily predicted from chemical theory (Ball 2001).

One of the most interesting questions in evolutionary biology is how new adaptations actually arise in organisms (Foote 1997; Gerhart and Kirschner 1997; Kirschner and Gerhart 2005; West-Eberhard 2003; Carroll 2005, 2006). Cartesian machine-based models fail when asked to deal with such questions because it is impossible to take an organism apart and actually learn or understand how it functions as a total integrated entity because taking an organism apart destroys its ability to function. It is impossible to tell how, or even if, an organism can respond to changes in its environment simply by examining its constituent parts. In ecological research this is described as "destructive sampling." The machine model is not useful in this context because it ignores the emergent properties that characterize living systems (Gould 2002; Kirschner and Gerhart 2005; Dorit 2007). Contemporary scientists who work at the cellular and molecular levels are figuring out that evolutionary change consists of two related phenomena: (1) conservation of essential processes at the cellular level, combined with (2) increasing diversity at the morphological and physiological level as a response to new or changing environmental conditions (Gerhart and Kirschner 1997; Kirschner and Gerhart 2005; Carroll 2005, 2006).

Difficulty in applying the Cartesian machine model arises from a variety of sources. To begin with, most evolutionary ecologists want to study adaptation, but they often assume that adaptation is something that took place in the past and is now complete. This means that all a biologist can do is describe how a feature might be adaptive; because it is assumed that evolutionary change is something that cannot be observed in the present. The Western system of knowledge and scientific investigation functions best at very small and very large scales—that is, where it is possible to transcend the obvious limitations of the human ability to directly observe phenomena, and investigators are most dependent on technology to provide data (Lewontin 2001; Alessa 2009). One problem, however, is that the more we tend to "proceed on hierarchic scales, the more metaphors we bring there from the scale we feel at home in. We end up transforming them into things we know from our experience" (Havel 1996).

Scientific theory can predict interactions between atoms or the orbits of planets in the solar system using such mechanistic approaches and their associated metaphors, but it is much less effective at understanding how animals interact within an ecosystem, or how individuals will interact within a population. Machine models function better at large and small scales because the properties of these systems are determined by small numbers of interacting forces, which are the basis of the physical sciences (e.g., gravity, chemical bonds). In biological systems, models derived from physical sciences are not applicable because analogues of mass, velocity, charge, etc., do not exist in any meaningful sense (Lewontin 2001). Without these simplifying concepts, modeling becomes much less realistic, and we enter the realm of relationships, where Western science is at its weakest.

A second problem that confounds the Cartesian machine metaphor is that organisms are internally heterogeneous in ways that are relevant to their functions (Lewontin 2001). As an example, circulation of blood in bodies is driven by the heart, which is made up of muscle, which is in turn made up of interacting molecules of actin and myosin, which are dependent for their action on generation of ATP. There also exists a horizontal multiplicity of functional pathways that define the function of parts in different ways—bones function as structural support for vertebrates but also serve as sites of muscle attachment, sources of calcium for metabolic processes such as lactation, and containers of bone marrow where new red blood cells are produced (Lewontin 2001).

Western science was slow to recognize that biological systems (whether organisms or ecosystems) are difficult to understand as parts and are best investigated as wholes. In other words, parts can be determined only after the "whole" has been defined (Lewontin 2001; Dorit 2007). Within the Western tradition, in the study of physical systems the "whole" can be defined by the space it occupies, and its behavior can be characterized by simple physical laws, whereas biological entities must be identified according to their function and the causal pathways that serve that function (i.e., relationships).

In contrast, Indigenous perspectives are most effective in observing and describing wholes, because they operate at the level of human perception and concentrate on functional relationships and co-evolutionary processes rather than structure (Barsh 2000; Alessa 2009). Indigenous intellectual and philosophical traditions largely ignore the structure of matter at the molecular level; instead they emphasize relationships between species or responses to environmental variation and the role of individual variation in population dynamics. The fundamental philosophical principles of Indigenous peoples, combined with their understanding of the unpredictability of the food sources upon which they depended, lead to intuitive understanding of the relationship between animal populations and the environmental conditions that shape the behavioral and anatomical responses of those nonhumans (Pierotti and Wildcat 1997a, 1997b). This principle has been elegantly outlined by an Indigenous scientist:

> "As you increase the level of space . . . or time . . . you increase the level of complexity, or how many things interact with each other at any given time. By the time you arrive at ecosystems (or populations), the interaction of organisms and their habitats, you have accumulated an enormous amount of complexity . . . As a consequence the scientific method and the Western approach to "understanding" is more tenuous and it is at this intersection of time and space that traditional

knowledge (TEK) is most apparent as another (alternative) approach."
Alessa (2009: 250; parenthetical elements inserted)

TEK-based approaches are fundamentally different from those taken by traditional Western science in its approach to understanding how organisms function. As an example, the American ecological tradition developed the idea of *r* and *K* selection from an assumption that the selective regime that governs reproductive strategies would be expected to change over the growth trajectory of an invading population (MacArthur 1972; Boyce 1984). MacArthur's basic idea was clever and insightful, because it assumed that the individuals initially invading an empty habitat would be selected to reproduce at a young age and a high rate. This would maximize their intrinsic rate of population growth (*r* in the basic mathematical model for population growth). Over time, as the population grew to a large enough size that it became established, factors such as competition between members of the same species for food and breeding sites, disease, and other density-dependent population interactions will emerge. The selective regime would then change, favoring adult survival and longevity over high rates of reproduction (MacArthur 1972; Boyce 1984).

In the tradition of the typical Western search for "laws of nature," Pianka (1970) attempted to generalize MacArthur's single population model in an attempt to explore the selective pressures that lead to the evolution of life history traits such as reproductive rate and lifespan across the animal kingdom. Pianka employed a comparative approach, assuming that different suites of selective pressures could be applied to a large number of species, and developed a dichotomous scheme that characterized different species (and higher taxonomic levels) as either *r* or *K* selected in terms of life history tactics, named for the mathematical notation used in the population growth equation. The designation of "*r* selected" can be thought of as occurring when population levels are low and the population growth rate is very high (think of *r* as loosely referring to reproductive output), whereas *K* refers to the carrying capacity. "*K* selected" species are those that show traits that allow them to persist at the high population densities that occur near the carrying capacity, such as superior competitive ability within their species.

Although it is not widely recognized, Pianka's approach was basically a life historical interpretation of the earlier dichotomy between "density independent" and "density dependent" regulation of populations (Andrewartha and Birch 1954; Lack 1954). In essence, Pianka transformed and simplified MacArthur's elegant model into yet another equilibrium-based approach to understanding population dynamics and reproductive tactics of animals and plants, which does not accurately reflect the real world.

Following logic established by Cole (1954), Pianka argued that *r*-selected organisms should show a set of traits that allow them to maximize the rate of reproduction, such as high numbers of offspring and short lifespan combined with early age of first reproduction. This combination of traits supposedly favored evolution of a semelparous life history where a species reproduces only one time during its lifetime and then dies (Pianka 1970; Stearns 1976). Pacific salmon of the genus *Onchorhyncus* discussed in the previous chapter in relation to First Salmon ceremonies are a classic example of semelparous organisms, however, unlike the predictions made by Pianka, salmon are not short-lived, nor do they have early reproduction or exceptionally high rates of fecundity.

Pianka argued that an opposite set of traits would characterize "*K*-selected" species, such as low birth rates, delayed onset of reproduction, long lifespan, and iteroparous life histories (which involve repeated bouts of reproduction). When a population grew to the point where there were so many individuals that food and other density-dependent factors became limiting (that is, it reached its environmental carrying capacity), population size was presumed to stabilize and to fluctuate only slightly around an equilibrium population size established by this carrying capacity. These assumptions of course, ignore the issue discussed in the previous two chapters, that is, that environments are not stable and that carrying capacity (*K*) is a misleading concept. According to the scenario developed by Pianka, a classic example of a *K*-selected species would be organisms such as bison or whales, who produce only one offspring every couple of years and have extended care of their young, but it is unclear that meaningful carrying capacities ever exist for such species (Pierotti 1991a, 1991b).

The resolution to this apparent problem was simple. Pianka's model simply failed to take into consideration several important characteristics of organisms and their environment, because it was based on a pair of simplifying assumptions: (1) that offspring are functionally equivalent to adults (yet another variation on the concept of a "typical" individual), so that a newborn has the same probability of survival as its parent, and (2) that environmental conditions are constant and do not fluctuate in an unpredictable fashion (Stearns 1976, 1992; May and McLean 2007).

The resolution of this issue came from the general recognition of two linked factors: (1) in most species, offspring survival is markedly lower than adult survival to the next opportunity for breeding, especially when environmental conditions are poor, and (2) environmental conditions fluctuate to such a degree that survival of offspring at times may actually be zero—that is, there would be no recruitment of new individuals to the population (Holgate 1967; Murphy 1968; Giesel 1976; Stearns 1976).

Mertz (1971a, 1971b) presented a more realistic alternative to the oversimplified view of populations that emerged from Lotka-Volterra equilibrium and stability-based modeling approaches. Following MacArthur's original logic, Mertz argued that most populations of long-lived organisms, such as trees and large mammals, birds, or fishes, undergo periodic rapid increases in numbers when conditions are unusually good. These "boom" periods account for phenomena such as "year-classes" in fish or "mast" years in trees. In between these intermittent but rapid increases, populations typically undergo extended periods when overall numbers slowly decline through attrition resulting from predation, disease, accidental death, starvation, etc. During periods of population decline, adults are much more important to the population's chances of survival than are juveniles, because they are physiologically better adapted to the conditions, have established territories and social groups, are usually larger and stronger, and have more knowledge and experience at survival. If environmental conditions deteriorate and the population continues to decline, the last surviving individuals will always be adults. The reality underlying this principle was demonstrated in a study of finches, *Geospiza fortis*, during a severe drought on the island of Daphne Major in the Galapagos. After two years of drought the only surviving finches were large-beaked adults (Grant and Grant 1989a, 1989b; Weiner 1995). These last surviving individuals were of high social status, and their body type allowed them to survive through a wide range of environmental conditions (see Chapter 9).

When conditions improve or even become very good again, these individuals will be the ones to reproduce and rebuild their population. An Indigenous gloss on this theme can be seen in the observation that "the strength of a tree comes not from growing thicker in the good years, but from staying alive during the bad, dry times" (Harney 1995). This quote from a Western Shoshone elder illustrates, as indicated in previous chapters, that Indigenous people had a clear understanding of how environmental conditions could fluctuate and create conditions where only the best-adapted individuals could survive.

There is an interesting conjunction here. If a society understands the reality of fluctuating environmental conditions, it is possible to link population dynamics to ecology in ways that reflect the real world, rather than the imagined stable, equilibrium-based world of European traditional ecology and economics. Early European ecologists tried to study single species without examining or understanding their relationship to other species, in what was called *autecology* (Kingsland 1985; Worster 1994). This tradition was replaced by *synecology*, which deals with the structure and dynamics of entire ecological communities.

POPULATION DYNAMICS IN FLUCTUATING ENVIRONMENTS: AN INDIGENOUS PERSPECTIVE

As described in Chapter 2, Indigenous understanding of the natural world emerged from conceiving of the living world as a network of relationships across communities that include humans. Because of this understanding based on relationships, Indigenous principles and insights are also superior at understanding links between systems that are often considered to be separate by the Western tradition. For example, scientists in Alaska could not understand why belukha whales, *Dephinapteras leucas*, did not enter certain river systems while at the same time they freely entered other rivers (Huntington and Myrmin 1996). Local Indigenous hunters attributed the avoidance of certain river systems by belukha to the presence of beavers, *Castor canadensis*. When the Western scientists expressed confusion at this interpretation, the Yupiaq explained that beavers built dams, which prevent salmon from reaching spawning areas in those rivers, whereas in rivers where beaver were absent, salmon were abundant. The belukha were drawn to salmon, not to rivers. As a consequence they avoid rivers where beavers and their dams are present (Huntington and Myrmin 1996).

What is interesting about this example is that Western scientists, both Russian and American, who were trained in marine biology could not conceive of the relationship between fresh water (beavers) and salt water (belukha) and organisms that bridged these systems (salmon). In contrast, the Indigenous Yupik saw the world in terms of relationships (see Kawagley 1995; also Barsh 2000; Henriksen 2009) and readily understood the links between these three species. This is a classic example of the compartmentalization of Western scientific knowledge (Nadasdy 2003: 123), and where the Indigenous way of seeing the world had a ready explanation based on connections for a phenomenon that perplexed the specialized thinking of Western science.

Despite the tendency on the part of Western science to denigrate Indigenous approaches and knowledge (Deloria 1995; preface), Indigenous traditions are based on understanding relationships between populations and fluctuating environmental conditions. A Kluane hunter felt that "the government could not effectively manage wildlife . . . because (it) has forestry experts, water experts, and mining experts . . . sheep biologists, moose biologists, wolf biologists, and bear biologists; and none of these people knows anything outside of their own specialty" (Nadasdy 2003: 123).

As with the beaver/belukha example described above, investigations that start from an Indigenous perspective may have a better chance of revealing population and ecological principles that underlie microevolutionary processes, as well as the evolutionary significance of cultural traditions in nonhuman societies, than does the Western scientific tradition. To revisit Mertz's (1971a, 1971b) idea that most populations are typically in a state of gradual decline, this idea clearly relates to Indigenous concepts of conservation that were set up to minimize the possibility of catastrophic declines and to keep prey populations sustainable and functional until the next period of increase. Not surprisingly, Mertz's insights concerning population dynamics were largely ignored by Western science, whereas Pianka's overly simplistic, equilibrium-based *r* and *K* selection became accepted as a fundamental principle of population and evolutionary ecology and continues to be taught in ecology courses (e.g., Ricklefs 1998) years after it was effectively demolished as a model because of its failure to make useful predictions (Stearns 1976; Boyce 1984).

As an example of how the Indigenous perspective views the functioning of and relationships between organisms and ecosystems, let us carefully examine the 1911 statement of Okute (Shooter), a Teton Lakota:

> Animals and plants are taught by Wakan Tanka [the Lakota creative force] what they are to do. Wakan Tanka teaches birds to make nests, yet the nests of all birds are not alike. Wakan Tanka gives them merely the outline. Some make better nests than others. . . . Some animals also take better care of their young than others. . . . All birds, even those of the same species, are not alike, it is the same with animals, and with human beings. The reason Wakan Tanka does not make two birds, or animals, or human beings exactly alike is because each is placed here to be an independent individual and to rely upon itself. . . . From my boyhood I have observed leaves, trees, and grass, and I have never found two alike. They may have a general likeness, but on examination I have found that they differ slightly. It is the same with animals. . . . It is the same with human beings, there is some place which is best adapted to each. . . . *An animal depends upon the natural conditions around it. If the buffalo were here today, I think they would be very different from the buffalo of the old days because all the natural conditions have changed. They would not find the same food, nor the same surroundings*. . . . We see the same change in our ponies. . . . It is the same with the Indians. (McLuhan 1971: 17–18 emphasis added)

In his 1859 book *The Origin of Species*, Charles Darwin laid out a set of conditions necessary for natural selection to occur that would lead to evolutionary change. These included 1) variation among individuals, and 2) that this variation should have an effect upon the ability of individuals to survive and reproduce. The major theme of Okute's statement demonstrates an awareness of variation among individuals. Okute also states that this variation leads some individuals to build better nests or take better care of their young, that is, to reproduce more successfully. Thus Okute has set up the conditions for the process of natural selection to occur in the context of traditional Lakota knowledge and spiritual teaching. It is also worth emphasizing that although Western anthropologists have tended to identify Wakan Tanka with the Western God, this force is more appropriately thought of as nature itself (Eastman 1980), and most Indigenous traditions identify creators with nature and the nonhuman (Pierotti and Wildcat 1997a, 1999b, 2000).

The relationship between Okute's statement and modern microevolutionary thinking becomes more explicit if we compare his statement, not to Darwin, but to Lewontin's contemporary (2001) restatement and updating of Darwin's principles: (1) "There are processes that produce variation among individuals within a population." (2) "Processes also exist that result in changes of the relative frequencies of the variants within a single generation." What is important is that Lewontin and Okute both emphasize the process of change, rather than the result of the change.

Another productive way of considering this issue has been discussed by the British evolutionary biologist Gabriel Dover: "It is the very transient nature of unique phenotypes, coupled to their biological trick of knowing how to reproduce, which makes them, through their specific exchanges with the local environment, the only determinants of the action of selection. Selection is here and now, not everywhere for all time" (Dover 2000: 63). Dover's description of the process of natural selection is similar in both its spirit and conclusion to Okute's statement about the different natural conditions that the buffalo would find today: "They would not find the same food, nor the same surroundings." Okute also assumes that changes in environmental conditions lead quickly to a discernable change in organisms, a concept that is not part of the equilibrium- and stability-based models of Western ecological thought.

Recent studies in evolutionary ecology have shown that organisms are capable of rapid evolutionary responses in relation to climate change and continue to change, which demonstrates that this is a dynamic process rather than a fixed outcome (Endler 1986; Weiner 1995; Hoffman and Parsons 1997; Reznick et al. 1997; West-Eberhard 2003). These organisms do not take their cues from the increase in temperatures associated with climate change, but from the resulting changes in seasonal events (Hoffman and Parsons 1997; Bradshaw and Holzapfel 2006). Usually this means that species are able to begin breeding earlier in the annual cycle or are changing migration patterns to take advantage of milder conditions at higher latitudes. This sort of response would be predicted by Indigenous science, because it emerges from human understanding of rapid environmental changes and how they affect temporal patterns in local ecosystems. "TEK is not merely a database. As long-term observers of ecosystems, Indigenous people discover associations in space and time" (Barsh 2000: 159; see also Brody 1982; Alessa 2009; Henriksen 2009). Indigenous people are cued to climate change, "when Indigenous people move through a familiar landscape, they augment their ability to monitor ecological processes by observing behavior of species with more acute senses" (Barsh 2000).

Okute is dealing with processes that he perceives as exemplified by Wakan Tanka, the Teton Lakota "Great Spirit," as manifested through the process of natural selection as defined by Lewontin (2001). One question frequently raised in studying Indigenous spirituality is whether the term *creator* should be considered as a verb connoting a process rather than as a noun connoting an entity. Keep in mind that *creator* is an English word, and that in Indigenous traditions the terms used that are translated as "god" or "creator" are much more process oriented. In many Indigenous languages verbs can express ideas that require entire sentences in European languages. Hence verbs are much more important (and much more abundant) relative to nouns in Indigenous languages, for example, in the

Mohawk language, there are approximately five verbs for every noun (B. Montgomery-Anderson, personal communication).

With regard to the concept of creator, perhaps the best analogue is to the English word *creation*, which simultaneously serves as a verb referring to the process of creation and as a noun describing the result of that process. In the case of the Lakota, *Wakan Tanka* can be envisioned as analogous to *creation* because it generates both the changed environmental conditions and the response to those changed conditions. It is clear that Okute associates variation among individuals within a species as being the result of changes in "the natural conditions around [them]." Of even greater interest is that Okute acknowledges that organisms change in response to changes in natural conditions over very short periods of time (i.e., over a few generations), which links his statements with contemporary studies on microevolutionary responses to changing environmental conditions.

TEK is based on an understanding of social (behavioral) relationships, both between individuals and among species (Barsh 2000; Alessa 2009), which are also the basis of ceremonies meant to enhance these relationships between humans and nonhumans. In contrast, Western science tends to understand and describe evolution in terms of changes in anatomy, focusing on the evolution of physical appearance, more than on changes in behavior and relationships. Indigenous thought, which emphasizes behavior and coevolution through the interactions or relationships among species, would be more attuned to identify rapid changes, even those that took place over just a few seasons.

Following this logic, I suggest that Okute's statements can be interpreted in part as recognition of rapid microevolutionary change over a few generations, a phenomenon completely unrecognized during Okute's lifetime, and only recently recognized by Western science (Endler 1986; Grant and Grant 1989a, 1995; Weiner 1995; Hoffman and Parsons 1997; Howard and Berlocher 1998). Until the 1980s, rapid microevolutionary change was considered by most Western scientists to be unobservable during human lifetimes. Only since investigators have been living closely with their study organisms and paying extremely close attention to them have these rapid and small changes been observed (examples in Endler 1986; Weiner 1995). What this means is that in essence, a few contemporary Western scientists have developed the equivalent of Indigenous skills with regard to recognition of individuals and observation of the roles these individuals play in their populations (see Chapter 6). These new-style Western scientists are in contrast to the entrenched Western approach based on average or typical individuals where inter-individual variation is not taken into account.

It is also important to keep in mind that the equilibrium- and stability-based population models used by ecologists are not based on the same assumptions as Darwinian evolution, which emphasizes variation among individuals. Population models, based on average individuals and factors like carrying capacity (*K*), which are characteristics of populations rather than of individuals, are hard to reconcile with observations of rapid evolutionary change. Successful evolutionary ecologists have had to transcend this barrier between ecological and evolutionary models and to take a new approach to studying organisms through careful observation in the field.

It is especially significant to note that Okute includes Indian people within his statement of the entities affected by the changes in conditions. This indicates an awareness that humans change in response to changes in environmental conditions in the same manner as do nonhumans. Okute would not be troubled by the idea of evolution happening not only in nonhumans but in humans as well. In contrast, the concept that humans and nonhumans are related remains controversial to this day in much of Western culture and philosophical tradition. Many major philosophers and scientists place a great deal of emphasis on defining and justifying the separation of humans from the remainder of the natural world (see below). Recent surveys have shown that less than half of the U.S. population believes in evolution, and less than 20 percent believe that humans and nonhumans are closely related (Scott 1997, Scott and Padian 1997; Miller et al. 2006).

The United States appears to be out of step with Western Europe and other Western-style nations in its attitudes toward evolution. Around 40 percent of Americans accept evolution in some form, and a similar percentage of Americans think that "evolution is absolutely false" (Miller et al. 2006). In contrast, nearly 80 percent of people in most Western European countries and Japan accept evolution. This level of ignorance in the American public is tied to many of the issues raised in this book. Lewontin characterizes the conflict as: "People believe a lot of nonsense about the world of phenomena, nonsense that is a consequence of a wrong way of thinking. The primary problem is not to provide the public with the knowledge of how far it is to the nearest star and what genes are made of, for that vast project is, in its entirety, hopeless. Rather, the problem is to get them to reject irrational and supernatural explanations of the world, the demons that exist only in their imaginations" (Lewontin 1997).

It must be kept in mind that many contemporary Indigenous Americans have become strongly influenced by Christianity and Capitalism and may now be counted among those who deny evolution. Some of my Indigenous students have initially considered themselves "creationists" until there was a class discussion on the differences between creation in the Indigenous and Western traditions.

"KEEPERS OF THE GAME"

As discussed above, the Western scientific tradition concentrates on how average individuals perform within a population and further assumes that such average individuals represent the optimal evolved state of the species under study. In consequence, Western science tends to largely ignore variation between and among adults in their roles as parents and the effects of this variation on breeding success. One reason for this failure to consider variation is because until recently the Western tradition assumed that all individuals of a species were basically similar in their ecology and behavior. For example, the Nobel Prize-winning ethologist Niko Tinbergen prevented his students from studying individual variation in gulls because he felt it would confound their studies of the adaptive bases of behavior (Burghardt 1997). Tinbergen's classic study on the behavior and ecology of gulls, *The Herring Gull's World* (1953), is a marvelous book, but it is built upon the assumption that describing the breeding behavior and ecology of an archetypical Herring Gull is sufficient to describe these phenomena for all members of the species.

In addition to emphasizing species-typical behavior, models in Western ecology tend to focus on population-level phenomena, rather than upon the individual as a member of the population. As a consequence, population-based models, such as the *r* and *K* models described above, emphasize "mean performance" within the population rather than individual variation in breeding performance (Lack 1968; May 1973; Pierotti 1991a). Such models work reasonably well to characterize rates of population growth, but they are not useful for characterizing the relationship between individual organisms and variation within the environmental conditions they face. As a result, population-based models cannot be used to study microevolutionary responses.

In a clear alternative to these average or mean-based population approaches in ecology, Indigenous tradition teaches that productivity in animal populations is determined and regulated by specific high-status individuals that are variously referred to as "keepers of the game" (C. Martin 1978), "animal masters" (Tanner 1979; Harrod 2000), or those "who cannot be killed" (Brown 1992). These "keepers" are individuals that "decide" whether or not their species will continue to be available for exploitation by humans. The relationship between these animal populations and human hunters depends upon the maintenance of good relationships between humans and these "keepers." If humans are greedy and hunt to excess, the "keepers" will withdraw their species from accessibility to humans (C. Martin 1978; Tanner 1979; Awiakta 1993).

"Keepers" are typically considered by Western anthropologists to be spirits, apparently somewhat akin to Platonic "ideals," or to be mythological constructs that serve a symbolic purpose. In contrast, we have the example of Kluane First Nations people expressing "concern over the practice of restricting hunters to shooting only full curl rams (mature rams eight years old or more). They argued that these older animals are especially important to the overall sheep population because of their role as teachers; it is from these mature rams that younger animals learn proper mating behavior . . . (and) more general survival strategies" (Nadasdy 2003: 127).

I have been told by non-Native anthropologists that it would be offensive to traditional Native Americans to think of "keepers of the game" as being anything other than spirits, yet this quote from Nadasdy shows that, in fact, this concept appears to be based on real social dynamics, albeit ones that are lacking from traditional educations in wildlife science. The Kluane people contend that, "Killing full curl rams has an impact on the population far in excess of the number of animals actually killed by sport hunters" (Nadasdy 2003: 127). Most Indigenous Americans I have asked about this disagree with the idea that "keepers of the game" are spirits, which makes me suspect that, as I discuss in Chapter 8, non-Natives are very protective of the images they create for Natives, even when these images may be the results of their own imaginations.

As part of my own research into variation in individual quality and reproductive performance of animals, I looked at whether the "keepers of the game" concept might be based upon actual experience with real animals. As argued above, Indigenous "legends" or stories about animals are metaphors that serve the same function as theoretical models in ecology. They present a generalized and oversimplified picture that is based upon a number of assumptions, some of which are based in knowledge and others of which are logical consequences of the structure of the model. Both models and stories are metaphoric structures intended to serve a heuristic purpose by imparting insights into how complex processes actually function. In the case of "keepers of the game," such stories may well be based upon actual experience with some extraordinary real animals, while also serving as heuristic devices that serve to teach Indigenous hunters to practice restraint and caution when hunting and to avoid taking individuals that might be responsible for a large percentage of the reproductive output of a population.

It is said by the Tsalagi (Cherokee) people that when a deer is killed, the "chief of the deer" comes to check the spot of the killing to find out if the hunter has behaved properly and respectfully. "When one of his own is slain, he instantly draws near and finding clotted blood on the leaves he bends low over the stain and asks 'Has the hunter prayed words of pardon for the life you gave for his own?' If the answer be no, then (the chief) goes—invisible, fleet as the wind—and tracks the blood to the hunter's home, where he swiftly pains and cripples his bones so he can never hunt again" (Awiakta 1993: 109). "No hunter who has regard for his health ever fails to ask pardon of the deer for taking its life" (Martin 2000: 36). Such stories represent instruction in how to hunt properly, so as not to overexploit key resources, and also teach that handling a killed deer carelessly can lead to crippling disease (e.g., Lyme disease from deer ticks).

As part of my own research into the evolutionary ecology of seabirds (see Chapter 6), I set out to examine the role of individual variation in reproductive output in regulating population dynamics, and also the potential impact of this variation on microevolutionary change. With colleagues I conducted a long-term (13-year) study of individual variation in lifetime breeding success in Western Gulls (*Larus occidentalis*) on Alcatraz Island (Annett and Pierotti 1999). We found a high level of variation in individual lifetime reproduction (0 to 29 offspring produced per pair over a lifetime), and also that most individuals were relatively unsuccessful (modal production of offspring among breeders was 0; see Figure 4.1 for details). Overall, we found that fewer than 10 percent of the birds that bred on Alcatraz were highly successful (i.e., only 9 of 124 pairs produced 20 or more offspring over their lifetimes; see Figure 4.1) and made a substantial contribution to the future population. Similar results have been obtained in all other long-term studies of breeding success in vertebrate animals when individuals have been identified and followed (Grant and Grant 1989b; see Clutton-Brock 1988 and Newton 1989 for numerous examples from birds and mammals).

Thus, a few individuals make a very substantial contribution to future generations. In general, the results of such studies have found consistently that about 10 percent of a breeding population can generate up to 85 percent of the recruits (defined as offspring that return to breed) in succeeding generations (Newton 1989).

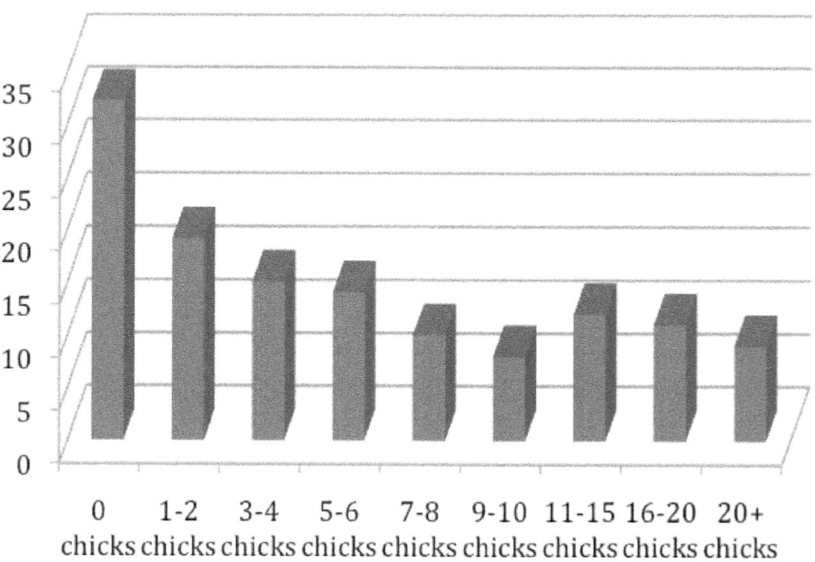

Figure 4.1 Distribution of lifetime reproductive success in pairs of Western Gulls, *Larus occidentalis* on Alcatraz Island, San Francisco, California. Y axis is number of individual pairs (one male and one female), X axis is the number of offspring that survived to fledging (independence from parents). Note that only nine of 124 pairs (8%) produce more than 20 surviving offspring (only four pairs produced more than 25 surviving offspring). These individuals are functional equivalents of "Keepers of the Game."

The theory of life history evolution in animals posits a trade-off between reproductive output (or reproductive effort) and survival, such that an individual who has high reproductive output lowers its own survival, or vice versa (Stearns 1992; Roff 1992). This trade-off is codified in the classic Lotka-Volterra equations, which assume that in a stable population where the number of individuals remains relatively constant, if the rate of survival (l_x) increases, then the rate of reproduction (m_x = reproductive output or maternity) must decline. This is because the rate of survival multiplied by the reproductive output (l_x times m_x) generates the typical rate of reproduction for an individual of a specific age and this is usually defined as 1.0. This assumed trade-off is referred to as the "cost of reproduction" (Bell 1980) and is a key component of the equilibrium theory of population dynamics.

Despite the predictions generated by the mathematical model, data from real populations show that the few high-quality individuals that show very high lifetime breeding performance also show high rates of survival. Therefore high-quality individuals do not show the hypothetical trade-off between survival and reproduction that is the basis of models for population dynamics and the evolution of life histories (Pierotti 1991a; Annett and Pierotti 1999). These individuals have both long life spans and high rates of reproduction, and contribute disproportionately to population growth (Figure 4.2).

Figure 4.2 A Keeper of the Game in the flesh: Western Gull Male #21 from Alcatraz. He produced a minimum of 27 fledged offspring from 32 eggs laid over 11 breeding seasons (Male #21 was a breeding adult at the beginning of our study, and we do not know how many offspring he produced prior to our work on Alcatraz).

If any individuals from this small number of successfully reproducing individuals were driven off by human activities, or even killed by a careless hunter, there could be a serious decline in local populations, as suggested by the Kluane people (above), which could occur as quickly as the next generation. These high-quality individuals show all of the attributes of the putatively legendary "keepers of the game"—that is, they live long, instruct their descendants on how to live effectively through cultural traditions, and their reproduction is essential to maintaining the health and size of the population. It is also important to emphasize, again as suggested by the Kluane, that these individuals are not easily replaceable if they are killed. They do not merely occupy high-quality territories; they are simply better than others of their kind at acquiring food, avoiding predation, and raising offspring.

Indigenous peoples, who paid very close attention to what was going on around them in the natural world and recognized many individuals in local nonhuman populations, were probably aware of the existence of such beings and integrated them into their combined knowledge and spiritual systems. Respecting and protecting such individuals would have been an effective way of minimizing the chance of local population crashes. In contrast, killing such individuals would mean loss of their reproductive output as well as the knowledge they would have held concerning distribution of food, water sources, and refuges. The consequence of killing or "offending" such individuals (showing lack of proper respect) could have led to shortages of key game species and starvation in people who did not show respect. As one First Nation's scholar describes it: "The consequences are not the ridicule of one's peers, or the failure to get research grants, they are sickness, suffering, and death (Alessa 2009: 250). Experiences like these would almost certainly have generated stories and traditions in which such individuals were given special status. Thus traditions arose that when disrespect was shown, the observed consequence would be that these individuals would take away the game, and the people would suffer (Martin 1978; Raban 1999; Bruchac 2003; Nadasdy 2003; Alessa 2009).

Population ecologists in the Western tradition seemed unaware that such high levels of variation in reproductive output and knowledge existed among animals until long-term studies of individuals began in the 1970s. Since there was no tradition of knowledge concerning such individuals, the first Europeans to hear such stories would have automatically considered such beings to be legends about spirit animals concocted by ignorant savages. An example of such thinking can be seen in the recent popular book *The Ecological Indian* by Shepard Krech (1999a; Chapter 8), in which the author insists that Indians believed in prey populations living under lakes or in caves during times of hardship without understanding the metaphoric component of Indigenous stories.

In reality, however, stories about "keepers of the game" are a more realistic description of actual population phenomena than many Western heuristic tools, such as MSY models, which form the basis for wildlife and fisheries population management. I use MSY models as an example because, as I argued in Chapter 1, stories in TEK are essentially discussions of how to manage natural resources from an Indigenous perspective, and MSY models are the Western scientific equivalent of such stories. MSY models are not effective tools for managing populations of large long-lived, low-fecundity species, such as marine mammals, and they are mostly ineffective even when applied to short-lived, high-fecundity species for which they were designed, such as sardines or anchovies (May 1973; Ricker 1977). Given this situation, it is particularly ironic that hunting by Europeans focuses on "trophy" animals, who are typically the largest and healthiest members of their populations and are probably those capable of making the largest contribution to future generations.

Even in the Western tradition there are stories concerning the existence of extraordinary individuals in natural populations of animals. In a long-term study of the George Reserve deer herd in Michigan, census drives were conducted annually in which all of the deer in this 50-acre fenced enclosure were systematically counted by biologists, with one exception. Within the enclosure there existed one very large buck who was never seen by the biologists. His existence was known because his tracks and shed antlers were found regularly, yet despite specific efforts to locate him, this male could not be found. Based on his size, and his probable high rank within the typical polygynous social system of deer, it was likely that this male was the father of most (possibly even all) of the deer present within the reserve (Dale McCullough, personal communication).

Similarly, in the long-term study of interactions between wolves and moose on Isle Royale in Lake Superior, there was one female wolf who survived for at least twelve years and bred for most of these. At one time it was estimated that over 75 percent of the wolves that made up the total population on Isle Royale were descendants of this single female. One characteristic that may have contributed to her success was that she was the best moose hunter ever seen among the Isle Royale wolves. On every occasion that she was observed to participate in an attack the moose was killed. There were occasions when she was observed to refuse to participate, and when attacked by other pack members these moose invariably escaped (Rolf Peterson, personal communication).

KEEPERS AS TEACHERS AND REPOSITORIES OF TRADITION

It seems obvious that the George Reserve buck and the Isle Royale female wolf possessed high levels of experience and knowledge. The female wolf knew which prey to take and which to ignore, and it is easy to imagine Indigenous peoples telling "stories" of such individuals. All studies that have examined the functional basis of variation in lifetime breeding success have found differences in diet choice or foraging ability between highly successful and unsuccessful individuals. In the classic study of red deer (*Cervus elaphus*) on Rhum, highly successful individuals control grazing areas where the forage is much higher in quality (Clutton-Brock 1988: 338). In Galapagos finches, the individuals with the greatest success held areas where high-quality food was more abundant (Grant and Grant 1989a, 1989b, 1995; Weiner 1995).

In our study of Western Gulls we found that birds with high lifetime output maintained a high-quality diet even during environmentally stressful periods such as the 1983 and 1992 El Niño events (Annett and Pierotti 1999). The differences in diet choice that we found in Western Gulls are almost certainly culturally transmitted, because male offspring, who remain with their parents after fledging, acquired the same diet as their fathers, whereas female offspring, who disperse after fledging, did not (Annett and Pierotti 1999). Thus, not only do a few individuals within a population show unusually long life spans combined with high rates of reproduction, but they are also probably repositories of specific knowledge that can be passed on to their descendants.

Evidence that older, experienced individuals serve as repositories of environmental knowledge that allow members of their social groups to survive has also been advanced for African elephants (*Loxodonta africana*) (Payne 2000, 2003; McComb et al. 2001) and sperm whales (*Physeter catodon*) (Vogel 1998; Whitehead 1998, 2003a, 2003b). As the Kluane people argued for older, more experienced rams (Nadasdy 2003), in these groups it has been argued that killing older, large individuals removes a tradition of knowledge upon which younger generations depend for development of skills and survival. This relates to the concept of exceptional individuals serving as "keepers," both of the game and of knowledge.

Another major difference between Indigenous and Western traditions of knowledge about the natural world is that until very recently the latter failed to recognize the importance of cultural traditions and family structures in nonhumans (Emlen 1995; de Waal 2001; Fragazy and Perry 2003; Laland and Galef 2009). Attitudes of lions toward human beings changed from tolerance to intolerance within only a few generations, which is much too little time for genetic factors to be involved, and strongly suggests that this behavioral change was the result of a loss of tradition (Marshall-Thomas 1994). This change resulted because the cultural transmission established by lions through interaction with local Juwasi peoples was ignored by Western park managers, who forcibly removed Indigenous people from the park—they felt that it was inappropriate for humans to exist in a national park. In the Western tradition humans are not considered part of the natural world (Dowie 2005) and are regularly expelled from areas designated as parks and preserves. As a result of these policies, lions, who may have peaceably interacted with humans for many thousands of years, lost the tradition of peaceful interaction and became aggressive toward humans.

Despite these examples and other extensive evidence of cultural traditions in nonhumans, there still exists a debate in Western science over the very existence of culture in nonhumans (Laland and Galef 2009). Investigators who argue for the existence of culture, or even cognition, in nonhumans are accused of anthropomorphism (Kennedy 1992; Emlen 1995; Budiansky 1998; Bekoff 2002). To Indigenous peoples, however, the existence of cultural traditions in nonhumans is an important component of TEK. It is assumed that animals have the ability of cognition (Anderson 1996), and that information is passed on to others about locations of good feeding areas or the most effective ways to hunt (Minta et al. 1992; Barsh n.d.). That animals are capable of such transmission is exemplified by many examples, such as high-quality gulls (Annett and Pierotti 1999) and wolves showing their descendants how to hunt effectively; high-quality elephants leading their social groups to areas for good feeding and where water is consistently available (Payne 2000; McComb et al. 2001); and high-quality chimpanzees teaching their offspring how to crack nuts or fish for termites (Whiten et al. 1999; de Waal 2001). There is extensive evidence of cultural transmission of foraging techniques in chimpanzees, *Pan troglodytes*, including local variation in the specific form of these techniques (Whiten et al. 1999; de Waal 2001), as well as the classic example of cultural transmission in Japanese macaques, *Macaca fuscata* (de Waal 2001; de Waal and Tyack 2003).

INCLUSION OF DIFFERENT TRADITIONS

This debate about anthropomorphic thinking and whether culture exists in nonhumans appears to result from the same materialistic Western philosophical tradition, developed by Aristotle and codified during the Enlightenment, that was designed to keep humans separate from the rest of nature (Mitchell et al. 1997). In fact, there remains a tradition in Western science in which scientists invest considerable intellectual energy to identify those criteria that make humans unique. This tradition appears to be yet another holdover from Cartesian philosophy, wherein it was believed that only humans had minds and were capable of thought, "the ghost in the machine" (Lewontin 2001; Bekoff 2000, 2002; Pennisi 2006). Over the years various criteria have been posited to separate humans from nonhumans, such as tool use, language, and culture. Interestingly, as our understanding of nonhumans and their ability to transcend these meaningless categories has increased, each of these criteria has been demonstrated to not be useful. The latest attempt seems to be over two issues. There is debate about whether or not nonhumans have a "theory of mind" based on their supposed inability to recognize whether or not members of their social group are capable of holding "false beliefs"—that another individual is thinking something wrong (Pennisi 2006).

The second issue is even more troubling from my perspective. It involves the fact that students of reciprocity and cooperation in humans often define their field in opposition to animal behavior. They argue that "Human society represents a huge anomaly in the animal world," supposedly because humans cooperate between nonrelatives, whereas animals restrict cooperation to close kin (de Waal 2009 describing work by Fehr and Fischbacher 2003 and Boyd 2006). The major difference really lies in that animal social groups are more likely to consist of close relatives than large human societies, but many animals show reciprocity and cooperation among non-relatives, including foraging associations and food sharing between nonrelatives, even so far as to involve different species (Pierotti 1988a, b; Rothstein and Pierotti 1989; de Waal 2009). There is also considerable evidence of animals adopting unrelated offspring (Pierotti 1989, 1991b).

Two ironies emerge from this tradition. First, why are Indigenous peoples considered to be primitive when they recognized the presence of cultural transmission, cooperation, and high-quality individuals in nonhumans for hundreds if not thousands of years? Second, and even more vexing given that the Western scientific establishment is bothered by the fact that most Americans, as well as many people in other parts of the world, do not accept the idea of evolution (Scott 1997; Scott and Padian 1997; Miller et al. 2006; Petto and Godfrey 2007), why is the question of what makes humans different from the rest of nature even being asked by the scientific community? If scientists insist upon the uniqueness of humans, they should not be surprised when nonscientists believe that humans were created separately from the rest of nature.

Looking at the evolution and ecology of nonhumans (and humans) from an Indigenous perspective could generate new ways to solve problems that have not yielded to the machine-based metaphors of standard Western approaches (Alessa 2009). As indicated above, in recent years approaches taken by Western scientists have begun to yield perspectives that converge on Indigenous traditional views of long standing, such as recognition of high variation in reproduction and survival, the beneficial role of predators in ecological systems, the benefits of controlled burning, and acknowledgement of the existence of cultural traditions in nonhumans. The Western scientific tradition has long assumed that there exists a single "truth" that is identifiable, and also that it has a monopoly on the approach to discovering such "truth." The discoveries described above suggest that there are several possible paths to take to arrive at an accurate portrait of the nature of interactions among individual organisms.

One example of how inclusion of individuals with different perspectives into biology has altered perspectives is that the increased number of female scientists in evolutionary biology has led to new discoveries and even new approaches to the study of social behavior and evolution (Gowaty 1996; Lawton 1996; Pierotti et al. 1996; Strum and Fedigan 2000). Female biologists have shown that there is more than one way of looking at "truth," and it is highly possible that more than one "truth" may exist. Therefore, inclusion of investigators from different cultural traditions may lead to even more "truths"—additional discoveries and even more dramatically different approaches to such studies. As an example of how this works, in Chapter 1 I mentioned how differences in the perceptions of Japanese investigators led them to numerous discoveries about primate behavior that were denigrated as "anthropomorphic" by Western investigators.

Indigenous people have a considerable body of empirically derived knowledge about the natural world in combination with a philosophical approach that is quite different from that found in any Western philosophical tradition (Mander 1991; Suzuki and Knudtson 1992; Barsh 2000; Pierotti and Wildcat 2000). Following traditions derived from Indigenous philosophy can allow humans to reconnect to the natural world at an experiential and spiritual level that provides much more meaningful connection than is often generated by the rather abstract and allegedly objective approaches of Western science. If the integration of Indigenous perspectives can be effectively achieved, it is possible that many of the foolish and unproductive arguments over evolution that characterize much of Western philosophical discourse will disappear. At the same time, such a merger could draw Indigenous peoples into a dialogue that could generate a truly integrated understanding of the relationship between ecological variation and evolutionary change.

5 Cultural and Biological Creation and the Concept of Relatedness

> In the Sioux story of creation [Wakan tanka] is not . . . conceived of in anthropomorphic fashion. . . . The enkindling warmth of the Sun entered into the bosom of our mother, the Earth, and forthwith she conceived and brought forth life, both vegetable and animal.
>
> Charles A. (Ohiyesa) Eastman, *Indian Boyhood*

In the previous chapter I introduced the debate over evolution and the reluctance of Americans to accept evolution compared with people in European countries (Miller et al. 2006). Two major scientific and philosophical issues are involved in this contemporary American debate over the empirical basis of evolutionary biology: *creation* and *relatedness*. These same two issues also lie at the heart of the differences between Indigenous and Western understandings of the nonhuman world and the way these cultural traditions interact with nonhumans. Resolution of these differences could go a long way toward increasing both Western scientific acceptance of TEK and increased understanding of evolution among the public as a whole.

Although little attention is paid to it in the media, the more important issue is *relatedness*—are human beings biologically related to the rest of the living world? The answer to this should be obvious, given the fact that all life forms share a wide range of physiological processes, including replication by means of DNA and RNA, the production of stored energy in the form of adenosine triphosphate (ATP) through the Krebs cycle and the electron transport chain located in the mitochondria, and the regulation of body form through the action of homeobox (Hox) genes (Kirschner and Gerhart 2005; Carroll 2005). If relatedness is accepted as a given, links to the rest of the living world are clear, and ethical obligations owed to these relatives by human beings is a logical philosophical development.

The problem seems to be that for some elements within American society acknowledgment of relatedness would be considered as tantamount to acceptance of evolution, which has been set up as oppositional to religious teachings, particularly with regard to the special creation of human beings as entities separate from the rest of nature. A leading evolutionary biologist has argued that what truly makes Charles Darwin's concept of evolution through natural selection controversial and the logical outcome of Darwin's theory of common descent was that the Darwinian model changed the position of humans from separate from nature to part of nature and subject to the same "natural laws" as the rest of nature (Mayr 1997: 182).

One major irony of denying relatedness between humans and other animals is that this idea underlies all of biomedical research. Nonhuman animal models are used as surrogates for humans in medical research. Use of animal models is based on the presumption that the physiological processes of nonhumans, especially our fellow mammals, are so similar to those of humans that it is possible to readily extrapolate from the responses of rats, mice, rabbits, dogs, cats, and pigs to those of humans. The evidence for both relatedness and the ability to make these extrapolations can be clearly demonstrated by the success of the drug treatments and therapies developed and tested on animal models that are currently saving human lives around the world. It is rarely argued by those who oppose the teaching of and belief in evolution that biomedical research should be stopped, because of the obvious benefits to humans, especially since these benefits accrue even to those who oppose evolution.

One odd situation that has arisen, however, is that some individuals who oppose evolution appear to be funding and allying themselves with "animal rights" activists in apparent efforts to stop biomedical research on animals. This is definitely one of those situations where politics makes strange bedfellows, with animal rights activists who are presumably on the political left, being funded by elements of the extreme right. As an example the Humane Society of the United States (HSUS), a presumably humanist organization given its title, has allied itself with right-wing politicians in a number of states in attempts to get legislation passed that opposes both animal research and results based on evolutionary biology. In one particular set of cases in which I have participated in Michigan, the Humane Society opposed legislation that acknowledged that wolves and domestic dogs are members of the same species, *Canis lupus*. This is contrary to current scientific understanding, since it has been recognized in the scientific literature since the mid-1990s that domestic dogs should be classified as *Canis lupus* rather than using the older Linnean name *Canis familiaris* to designate the domesticated form as a separate taxon (Roeder and Wilson 1995). As a corollary to this effort, HSUS and the American Veterinary Medical Association have opposed the use of rabies vaccines for use in wolves and even in crosses between dogs and wolves, on the basis of the argument that these are different "species," even though these organisms are both genetically and physiologically indistinguishable. What goes unrecognized and unacknowledged is that this insistence that dogs and wolves are "separate species" is basically a creationist argument, which assumes that when organisms are considered to be "distinct species" they can then be assumed to be the result of distinct creation events rather than evolution through common descent.

In contrast, Indigenous Americans have recognized the identity and relatedness of dogs and wolves for millennia. For example, the Nuhmuhnuh would not eat dogs because they considered them to be the brother of their creator, Wolf (Wallace and Hoebel 1948). The Tsistsistas used domesticated wolves/dogs as their pack animals and to pull travois before the arrival of Europeans (Schlesier 1987; H. Mann, M. Little Thunder, personal communications). Osages even used the same word, *Shon'-ge*, for both wolves and dogs (LaFlesche 1995: 132). In general, Indigenous peoples accept relatedness as a given and employ the concept of the relatedness of all life forms as a major principle of their belief systems.

The second issue in the debate over evolutionary biology is the concept of *creation* and the meaning of this term to people of various cultural traditions. The term *creation* has become politically charged in American scientific circles in recent years, presumably because the term is associated with the "creationist" movement in American religious tradition, which has a history of being anti-science, especially through its opposition to the teaching of evolution in public schools (Scott 1997; Scott and Padian 1997; Petto and Godfrey 2007). The "creationist" movement contends that all life forms, but especially human beings, resulted from "creation" events directed by an anthropomorphic God who serves as the "creator" (Petto and Godfrey 2007). The logic that appears to underlie this Western religious tradition is that human beings are separate from the rest of nature, which is being taken to its logical extreme by fundamentalist Christians (Mayr 1997; Coates 1998; Pierotti and Wildcat 1997a, 1997b, 2000).

Confusion exists in both directions. Bias against the use of the term *creation* has become so extreme in scientific circles that when I was writing an essay about the belief systems of Indigenous Americans for a special issue of the scientific journal *Ecological Applications* (Pierotti and Wildcat 2000), the editor insisted that we remove any reference to the terms *creation* and *creators* from our essay, on the grounds that this was likely to lead readers into thinking we were "creationists." Ironically, the point I was trying to make in that essay is that *creation* means different things to different cultures, and that the term *creation* should not be conceded to the religious right because every Indigenous culture has its own "creation" story and concept of a "creator."

Indiscriminate and careless use of the term *creation* can lead to confusion, however, even among Indigenous peoples. During a discussion at Haskell Indian Nations University in the mid-1990s, I had a superb Indigenous student from the northern Plains state that even though he was majoring in biology, he believed himself to be a "creationist" because he accepted the creation story of his people. Subsequent discussion revealed that this student believed strongly in the relatedness of human beings to other life forms, as well as believing that organisms changed in response to changing environmental conditions. The problem was that this student assumed that Western fundamentalist Christian views were equivalent to the creation stories of Indigenous American peoples because both used the term *creation*, even though the concepts involved are very different. This exchange illustrates the fundamental confusion that can result from conceding the term *creation* to religious fundamentalists.

What often goes unrecognized is that the concept of "creation" typically has a very different meaning in the spiritual traditions and knowledge bases of Indigenous peoples (Kidwell et al. 2001). As emphasized in the previous chapter, "creation" can be thought of primarily as a verb describing a process that is ongoing. This is illustrated in the epigraph to the present chapter. Many Indigenous peoples do not think of their "creator figure" as an anthropomorphic figure, but as a process involving an interaction between the earth and the sun that led to life.

In Western monotheistic religious traditions the concept of creation appears to be related to the beginning of time itself, which is a Western philosophical construct with no empirical meaning. The Western intellectual tradition is temporally oriented (Deloria 1992) and places excessive emphasis on dates that are often defined in an arbitrary manner. For example, what does it mean that we are in the year 2010, or even the twenty-first century? These numbers are supposedly based on a system determined by dating from the birth year of Jesus Christ, hence the use of the terms BC (before Christ) and AD (*anno Domini*, in the year of the Lord). Most contemporary cultures in the world use this dating system regardless of whether or not they are Christian in their beliefs, yet these dates have no real meaning. Much of the panic over the arrival of the new millennium assumed that the year 2000 had meaningful significance to the fate of the world; however, nothing happened, and nothing was likely to happen, because the date had no empirical meaning in terms of the physical history of the earth.

Similarly, the beginning of time has multiple interpretations based upon a variety of understandings and philosophical perspectives. Fundamentalist Christians believe that the earth is less than 10,000 years old, and that the age of the earth can be dated by counting backward from the birth of Christ through the ages of prophets and various biblical figures until they reach Adam and Eve, at which point they add a week for the "seven days of creation" (Petto and Godfrey 2007). Young Earth creationists believe that the world was created in six literal (twenty-four-hour) days, and they do not seem to consider that anything exists in the universe except the earth and perhaps the solar system. I presume they think that the sun must also be less than 10,000 years old, although this point is never addressed in their writings and teachings.

In contrast, the prevailing view in Western science is that the universe began several billion years ago. Since there is no usable date attached to the events such as the "Big Bang," there can be no way of assigning a date under this system. This may explain why even scientists adhere to the BC/AD dating system even though this commits them to acceptance of a Christian tradition that many do not follow (many people today use BP, which means "Before Present" as a substitute). In contrast to Christianity, other cultures such as Judaism, Islam, Buddhism, and Taoism employ different dating systems based on their own understanding of the timing of significant events in their history.

One major debate in the creationist/evolutionist controversy, especially in my current state of residence, Kansas, is the timing of the beginning of life. Creationists seem to believe that life and time began with the Garden of Eden, and the seven previous days their god spent in creating the world. Ironically, the origin of life is no longer considered to be an important question among evolutionary biologists, even though this was undoubtedly an important event, or more likely a series of events, that is, an ongoing process (Shermer 2006). Despite the beliefs of creationists, Charles Darwin never discussed the origin of life in any of his writings (Ruse 2008). Darwin was much more concerned with how life changed once it existed than with its primordial origins. As a contemporary scholar has said, "Darwin started in the middle; he cautiously refrained from pushing his own published thinking back to the beginning—the ultimate origin of life and its preconditions" (Dennett 1995: 149).

Contemporary textbooks in evolutionary biology barely even address the issue of the origin of life and typically begin their discussion of life with the first findings of cells in the fossil record during the Precambrian period, some three billion years ago (e.g., Ridley 2004). Evolution is based on the idea that once life existed, life forms responded to variation in the environment that allowed some forms to survive and reproduce themselves more efficiently than did other similar life forms (Darwin 1859; Gould 2002; Kirschner and Gerhart 2005; Carroll 2006). The difference in assumptions about the age of the earth is the essential issue in this debate; after all, if people believe that the earth is less than 10,000 years old, the concept of three billion years, or even several million years, can be problematic to their concept of both the world and their place in it.

"INDIGENOUS CREATIONISM"

Unfortunately, for Indigenous people this issue has been clouded by some illogical arguments that have been presented by several Indigenous scholars in recent years, leading to confusion about evolution among both Indigenous scholars and students. There is a troubling tendency for Indigenous scholars in nonscientific fields to argue that they do not believe in evolutionary change (e.g., Deloria 2002; Atleo 2004; Tinker 2004). These scholars all appear to argue that evolution is not relevant to modern Native Americans, which inhibits the ability of scientifically trained Indigenous scholars to posit any alternative views.

The arguments presented by Vine Deloria Jr. are especially troubling in that his statements are almost indistinguishable from those of a Young Earth creationist of European ancestry—that Mother Earth is so new that she came into existence around the time that humans arose, including Native American peoples (Deloria 2002; Shermer 2006; Petto and Godfrey 2007). From my point of view, this argument of Deloria's amounts to an insult to our planet, since the earth has been in existence so long that humans are a very new species. Deloria appears to be inconsistent about his views concerning the age of the earth. He has written that human beings were "the last creatures to be created and the youngest of the living families" (Deloria 1999b), which contradicts much of what he subsequently said in his 2002 book *Evolution, Creationism, and Other Modern Myths*, and Deloria's inconsistent but strongly argued views need clarification.

Deloria's (2002) basic contention seems to be that evolution is equivalent to Western science, and that science functions as a new religion. One of the more problematical statements made by Deloria is that now that science has replaced religion, "Nature" has replaced "God." This argument specifically contradicts earlier statements made by Deloria, such as:

> No element in of life [*sic*] can go unattached from human society. Thus lands are given special status because they form a motherhood relationship with the peoples who live on them. . . . *The true meaning of the motherhood of the land is that, like a mother, it shapes and teaches our species*, and according to the peculiarity of the area, produces certain basic forms of personality and social identity which could not be produced in any other way. . . . With respect to other life forms this attitude manifests itself in what one could call "kinship" cycles of responsibility that exist between our species and the other species. . . . This transformation brings out a dimension of life common to Indians, but unique and unsuspected by non-Indians. (Deloria 1999a: 131, emphasis added)

If this is what Deloria truly believes, he is close to equating Nature (or land) with the process of creation. To be intellectually and philosophically consistent he should embrace the concept that God is equivalent to Nature. If we examine the quote that begins this chapter from Charles Eastman (Ohiyesa), who like Deloria is a mixed-blood Sioux, we can see two things. First, Eastman appears to be less colonized in his thinking than does Deloria, in that he does not use metaphors and arguments from fundamentalist Christianity, as did Deloria (2002). Second, Deloria's 1999 description of the land being "like a mother" is very similar at its heart to Eastman's view of the nature of the Dakota creator *Wakan Tanka*. Given the philosophical and spiritual traditions of Deloria's own people, and of his own earlier writing, it seems to be appropriate to consider Nature as equivalent to the "creator" no matter what Deloria said in 2002 (Deloria and his writings are discussed in greater detail in Chapter 9).

Richard Atleo (2004) attempts to derive an Indigenous theory of how the natural world functions based on the traditions of his Nuu-chah-nulth people. His theory is based on the Nuu-chah-nulth concept of *heshook-ish tsawalk* or "everything is one." This concept seems conceptually related to the idea that "all things are connected," which I argued was the basis of

Indigenous concepts of community in Chapter 3. Atleo uses ideas borrowed from quantum physics, including Heisenberg's uncertainty principle, in an effort to link the physical world with the spiritual (xii–xvi), an approach similar to that taken by Calvin Martin in an attempt to provide an empirical basis for the transformations of shape and form that are often described in Indigenous stories (Martin 1999).

I admire much of Atleo's effort, and I recommend it to any reader interested in trying to understand an Indigenous worldview. I cannot understand, however, why he chooses to make the statement that "whereas the theory of evolution holds that life evolved from simple to complex, from primitive species to more advanced species, the theory I present . . . holds, in keeping with traditional origin stories, that life did not evolve, but began as complexity" (Atleo 2004: xix). Like Darwin, Atleo is starting in the middle, and he seems to derive his ideas about evolution from Plato in that he invokes concepts analogous to the primacy of perfect forms and seems to argue that creation happened only once and that all beings have existed unchanged since that time. Atleo appears to function as an extreme rationalist, in the sense of the view of the Greek Parmenedes who maintained, "whatever the senses might say, the very notion of change involved a contradiction, and so reason demanded that reality be entirely void of change" (Honderich 2005: 783).

This argument is more sophisticated than that employed by Deloria, in that Atleo clearly acknowledges the existence of life forms prior to humans and relatedness between human beings and the rest of the natural world. He states that "in the Nuu-chah-nulth worldview every life form is of one *thli-muhk-sti* (spirit), of one essence" (Atleo 2004: 61), and that "the source of creation . . . established the life principle of *like producing like*" (62, emphasis in original).

Atleo's argument is similar in many ways to arguments advanced by Sir Richard Owen and Baron Georges Cuvier in the early part of the nineteenth century about the immutability of living species. This is puzzling because Atleo also states: "one of the major themes of creation is transformation or change. Things and life forms do not just hold still. Existence appears to be dynamic rather than fixed" (Atleo 2004: 63). This seems to accept the idea of change, yet Atleo goes on to argue, "In contrast [to the Nuu-chah-nulth worldview], the predominant scientific worldview based in Darwinian theory postulates that biodiversity evolved through a random process of evolution. . . . This means that some life forms cannot be fit to survive" (127). To Atleo, accepting natural selection seems to imply that the Nuu-chah-nulth creator *Qua-ootz* has not made all living beings equal, and this violates the principle of *heshook-ish tsawalk*.

This argument does not follow its own internal logic. Atleo says that "one of the major themes of creation is transformation," but this change cannot be related to differences in the ability to survive or reproduce as argued by both Darwin and Okute (Chapter 4). I suspect that Atleo may be conflating *Qua-cotz* with the anthropomorphic Christian God and arguing that the perfect (Platonic) forms are all equal, which represents a very convoluted approach to understanding the nature of variable living organisms in a variable world. A world that is highly variable is always going to produce the "random" effects that he rejects. Such a world also establishes the basis for Indigenous spiritual systems that are based on coming to terms with a changeable world (Anderson 1996; Raban 1999; Chapters 2 and 3). Ironically, if Atleo did not make these inconsistent statements about his understanding of evolution, his ideas could be used to argue that the Nuu-chah-nulth worldview based on *heshook-ish tsawalk* incorporates ideas not very different than modern evolutionary theory. Certainly his worldview seems to be based on the idea of relatedness between different forms of life.

The most unusual, yet most straightforward, statement about evolution by an Indigenous scholar comes from George "Tink" Tinker, an important scholar of Indigenous religious studies: "As an American Indian . . . I must confess that I am not yet a believer in the 'fact' of evolution. I do not believe that we Osages evolved from monkeys. . . . At the same time, Indians should continue to have deep concern over the Euro-Western proclivity for understanding *descent as a category of ascendancy* leading to anthropocentric privileging of the human mind. American Indians are deeply aware of our part in another family tree entirely—one predicated on interrelationship rather than on descent or hierarchy of any kind. As Osages, our closest living relatives are our sisters the buffalo and our sister corn" (Tinker 2004: 108, emphasis added).

Like Atleo, Tinker appears to believe strongly in relatedness but has difficulty in coming to terms with the meaning of creation. His statement is fascinating, both in its failure to address actual evolutionary theory (i.e., his idea that humans evolved directly from "monkeys," which is not part of evolutionary theory, but instead a common popular caricature of evolutionary thought) and in his apparent acceptance of actual evolutionary principles, that is, that humans are closely related to other nonhuman species.

Part of Tinker's problem seems to result from a popular misunderstanding of Darwin's reason for using *descent* in the title of his seminal work on evolutionary change resulting from natural selection. Darwin used the term *descent* in his second book, *The Descent of Man and Selection in Relation to Sex* (1872), as an ironic comment on the Scala Naturae—the idea that life was hierarchical and creation led directly from simple life forms through other vertebrates to humans, and also to invert the idea of the *Ascent of Man*, a concept that underlay Victorian concepts of society and the progress of mankind (Worster 1994). Darwin's intent was to point out that, as believed by most Indigenous traditions, humans were part of nature, which is why he used the term *descent* to imply that humanity was not in fact the peak of creation as his scientific predecessors had argued. Even though Tinker fails to realize it, it is those who caricature and simplify

Darwin who he opposes, rather than Darwin himself. The problem is, like both Deloria and Atleo, Tinker does not cite the actual literature on contemporary evolutionary biology but rather uses popular misunderstandings of evolutionary theory. If these scholars were to address the scientists working in evolutionary biology directly, they would learn that the idea of ascent or descent is no longer in favor and that the idea of hierarchy has been dropped, even from the field of modern systematics (Margulis 1998; Carroll 2005; Kirschner and Gerhart 2005).

CREATORS AND "CREATION"

Indigenous concepts of creators and creation are different from those of the Western scholarly tradition, which, despite contemporary debate over religion, has deep roots in the Christian religious tradition (Coates 1998; Cajete 2000). Indigenous concepts are actually much closer to current evolutionary thinking than is the Western monotheistic tradition. Indigenous people see the profound variety of living beings not as the result of ordered variations in the mind of a transcendental super being, but as society resulting from the interdependence of spirits that binds one life form to all the others.

What is unfortunate is that none of the Indigenous scholars discussed above invoke modern evolutionary theory to address its implications for Indigenous spiritual thinking. The idea that humans were one of the last life forms to evolve is a principle of both evolutionary theory and of Indigenous spiritual tradition as described in the writings of the very Indigenous scholars (Deloria 1999b; Tinker 2004) who regard themselves as critics of evolutionary concepts. The ultimate irony is that both Deloria and Tinker (possibly Atleo as well) are displaying mindsets that result from colonization—they respond like fundamentalist Christians in their writing about evolution. This is particularly ironic in the case of Deloria, who has written, "I have in my lifetime concluded that Christianity is the chief evil ever to have been loosed on the planet" (Deloria 1999b: 146), yet he invokes Christian-derived arguments to criticize evolutionary ideas. As a result, their arguments assume a form similar to that seen in the confrontation between fundamentalist Christians and "popular culture's" misconceptions about evolution, rather than what would be expected in a dialog between a scholar speaking from an Indigenous view with a modern scientist working in the field of evolutionary biology.

As a response to the comments made by Deloria, Atleo, and Tinker, I would like to cite other Indigenous writers. "It's always been a journey. It's a journey about the myth of creation. It's not a static, monotheistic 'God dropped you or created you from clay!' It's the journey. In most Native communities it's a journey to other life forms—animals, birds, other zones of consciousness. These are all journeys, creations. There is no static creation, no one moment like on the delivery table" (Gerald Vizenor, cited in Eigenbrod 2005: 172). Even more to the point is the statement by the Canadian Aboriginal (Stoh:lo) writer Lee Maracle, "If you have only yourself as a start and end point, life becomes a pretence at continuum" (cited in Eigenbrod 2005: 61). Two writers from Laguna Pueblo have also weighed in on this topic. First, "For the American Indian the ability of all creatures to share in the process of ongoing creation makes all things sacred" (Allen 1986), and second, "The ancient Pueblo people called the earth the Mother Creator of all things in this world" (Silko 1996).

Deloria seems to accept relatedness if we rely on his earlier writings (Deloria 1990, 1992, 1999a, 1999b), but he appears to have the same problems understanding creation that I discussed earlier in regard to equating "creator" with the Christian God rather than with "creation." Thus when Deloria needs an expert on science, he cites theologians like Paul Tillich or Hans Kuhn. In the same chapter he describes Steven J. Gould as "one of the most outspoken evolutionists" (Deloria 2002: 2), while later he describes Gould's idea of punctuated equilibria as being equivalent to creation and Gould as the "creationist's best friend" (16–18). The key problem seems to be that Deloria simply did not understand punctuated equilibrium as it pertains to modern evolutionary theory, as opposed to catastrophism.

Part of this problem may have arisen because instead of citing Eldredge and Gould's original paper on this topic (1972), Deloria cited books written for popular audiences by both Gould and Eldredge (Gould 1977, 1999; Eldredge 1985) and appears to make the same mistakes in interpretation of these ideas that are now commonly found in the dominant culture. The idea of punctuated equilibria is based on the observation that in the fossil record, organisms appear to change form rapidly, especially during periods of major environmental change. In between such major changes the fossil record indicates that organisms do not appear to change dramatically, although they often change in small ways, for periods sometimes lasting up to millions of years. The punctuated equilibrium argument is meant to apply to the anatomical changes that can be seen in the fossil record, rather than to the smaller changes in physiology, behavior, soft tissues (e.g., color of skin, feathers, or fur) and ecology, which are the stuff of microevolutionary change discussed in the previous chapter. Gould and Eldredge are paleontologists who study evolutionary changes in the fossil record that occurred in the past, not evolutionary biologists studying microevolutionary changes ongoing today.

Turning to microevolution, Deloria represents it in the following way. In discussing a debate between criminal lawyer and opponent of evolution Philip Johnson and an unnamed scientist, Deloria describes the scientist (whom he refers to as an "evolutionist," thus echoing "creationist" language) as citing "the old Darwinian finch story in which birds having slight differences in their beaks were understood by Darwin as evidence of evolution . . . [but he] failed to note that more recent studies have shown that the finches' beaks change with the amount of rainfall each year and that no evolutionary change is taking place" (Deloria 2002: 21).

If we look at the history of Darwin's work on the Galapagos we find that this is a more complex and interesting story than is usually discussed in the popular press. Darwin's ideas changed in fundamental ways because of his careful observations of communities of plants and animals during his voyage to the New World. Darwin looked directly at Nature and Nature looked directly back at him, changing his life forever. Not yet a scientist, Darwin was deeply impacted by his increased understanding of the relationships and connections between organisms.

In contrast to the popular story, Darwin did not assume that the Galapagos finches showed evidence of evolution because he did not initially realize that these birds were members of different species. Subsequent work by others, who referred to these birds as "Darwin's finches," revealed the existence of at least eight distinct species of finch, all of which appear to be consistently identifiable species (Weiner 1995). The studies of this group of finches that provided convincing evidence of microevolutionary change were not conducted until the 1970s and 1980s, more than a century after Darwin's time. The more recent studies to which Deloria refers are the classic research conducted by Rosemary and Peter Grant (Boag and Grant 1981; Grant and Grant 1989a, 1989b, 1995), which demonstrated beyond doubt that natural selection, which Darwin had assumed worked too slowly to be observed, was taking place over very short periods of time. Substantial evolutionary change was, in fact, observed in empirical research conducted by the Grants (Grant and Grant 1989a, 1989b, 1995). Finches with larger beaks were favored during a drought. This pattern partially reversed itself during a subsequent El Niño event (Grant and Grant 1989b), which is apparently to what Deloria refers in his comment about changing "with the amount of annual rainfall."

Rather than no evolutionary change taking place as claimed by Deloria, in the Grants' studies evolutionary change was observed to occur constantly (Grant and Grant 1989a, 1989b, 1995). Deloria appears unaware of the Anishinaabe writer Gerald Vizenor's statement on this point, "Creation myths are not time bound, the creation takes place in the telling, in present tense metaphors" (Vizenor 1981: xii), which is highly ironic given Deloria's (1992) ideas concerning temporally oriented thinking. The Grant's study established that microevolutionary change was real, and could be convincingly demonstrated during relatively short-term studies (see Endler 1986, or the Grants own 1989 work for technical accounts and Weiner 1995 for a popular account).

Ironically, studies such as this one by the Grants actually derive from careful attempts to study nonhumans as individuals trying to survive in unpredictable environments. Such studies converge on the Indigenous understanding of ecology and interactions and relationships between living beings and the natural world. Deloria could have used the Grants' studies to argue that Indigenous approaches were superior to those employed by traditional Western science. Instead, Deloria and Atleo position themselves in opposition to evolutionary biologists and seem determined to establish that "origin stories" are literally true. Such adherence to foundational principles as literal truth is a form of fundamentalism. As is the case with creationists of European ancestry, the arguments made by Atleo, Deloria, and Tinker seem incapable of imagining a world that predates humans.

CREATION AS A SPATIALLY-ORIENTED CONCEPT

The concept of "creation" or "origin" in Indigenous cultures seems to refer to a series of events, not located in specific periods of time, but instead to a particular physical location or place, where the people in question came to recognize themselves as a distinct cultural entity. My Arikira/Hidatsa colleague Michael Yellow Bird has argued that creation or origin stories are about survival of peoples through difficult times, such as major environmental change (personal communication). The message of such accounts is that if our ancestors could survive through such times, then surely we can survive through the times we are currently experiencing, which are much less harmful.

This clearly relates to the idea that Indigenous cultures were attuned to variability in the environment, and that such cultures were capable of making major adjustments in life ways in order to accommodate environmental changes. Laguna Pueblo scholar Paula Gunn Allen has written, "Tribal systems are static in that all movement is related to all other movement . . . they are not static in the sense that they do not allow or accept change. Even a cursory examination of tribal systems will show that all have undergone massive changes while retaining those characteristics of outlook and experience that are the bedrock of tribal life" (Allen 1986: 63).

An analogous version of this interpretation can be found in the Jungian tradition, where it is assumed that "creation myths . . . represent unconscious and pre-conscious processes which describe not the origin of the cosmos, but the origin of man's [*sic*] awareness of the world" (von Franz 1995: 5). In either interpretation it is clear that the concept of "creation" should be applied to the origin of human awareness, or to the beginning of a new cultural tradition, which is fundamentally the same thing.

It is worth noting that the origin of a new cultural tradition in response to a change in specific environmental conditions is analogous to the formation of new species in relation to specific selective pressures imposed by environmental conditions. Formation of new species is typically assumed to be the result of drastic new changes in environmental conditions (Mayr 1961, 1997; Coyne and Orr 2004; Kirschner and Gerhart 2005). As with species, cultures have been shown to exhibit variation, competition, inheritance, and the accumulation of successful modifications over time (Mesoudi et al.

2004). In the case of cultural changes, adaptation results from behavioral changes and can occur extremely rapidly, because cultural evolution can function in a Lamarckian as well as a Darwinian manner. This emphasis on behavior fits well within the idea that Indigenous knowledge and TEK are based more on behavioral and ecological observations than on genetics and morphology (Barsh 2000).

A wonderful example of the evolutionary nature of creation stories can be seen in the origin story of the Navajo (Dine'e). This story is very complex and sophisticated, with insights that converge strongly on evolutionary ideas. For example, the first, or black world, contains only invertebrates, "crawling creatures, slugs, and larvae . . . and not even the insects had wings" (Newcomb 1990: 1; J. Peshlakai, personal communication). Life then moved to the second, or blue, world, which was dominated by birds, who went to war against the insects, pursuing them into the third, or yellow, world. The yellow world contained mammals, such as chipmunk, packrat, coyote, and mountain lion, and also humans. Each transition from one world to another in this story is prompted by extreme environmental stress. Most interesting is the fourth world, to which the people are forced to move by food shortages and extreme competition among the different peoples/species (Newcomb 1990: 5).

As in the creation stories of other Indigenous peoples, the Dine'e consider the other species to be equivalent to themselves. When the Dine'e arrive in the fourth world, they find the Pueblo and Comanche peoples already there. This is almost certainly an accurate representation of the actual experience the Dine'e underwent as they moved into the Southwest within the last millennium. This also clearly establishes that this is an origin story of a people and their cultural tradition, rather than a story about the beginning of life. Ironically, the Euro-American chronicler of this origin story does not consider this to be a "creation story," because "no mention is made of a creator; nor does a continuous plan of action occur" (Newcomb 1990: xxiv). To me, the Dine'e origin myth is an insightful story cycle, which deals in a sophisticated and nuanced fashion with the ecology of local species and the changes they undergo as their world changes around them. Several of these species help instruct and guide the humans to their new way of life as the Dine'e learn how to survive and then thrive in the new worlds they experience.

Another important difference can be observed in the nature of these "creators" (when they are considered to be present and active participants) described by Western and Indigenous creation or origin stories (i.e., the entities presumed to be responsible for the beginnings of important cultural traditions). In most Western belief systems identified "creators" tend to be human, or human in form. This trend is clearly true of the three major monotheistic Western religions: Judaism, Christianity, and Islam, where Jehovah (Yahweh), God the Father, and Allah, respectively, are all clearly conceived of as human in appearance and male in gender.

By contrast, within Indigenous belief systems creators are typically nonhuman entities who are represented by a species of animal or plant that is considered important to the local ecological community (Pierotti and Wildcat 1997a). This tradition reinforces the idea of both relatedness and connectedness because significant nonhuman species are considered to be the originators of cultural tradition and sometimes of human beings themselves. This is exemplified by the treatment of Raven as the creator figure in cultural traditions of the Pacific Northwest and Alaska (Nelson 1983; Anderson 1996); by the treatment of Wolf and Coyote as creator and trickster figures, respectively, in the Plains and the intermountain West (Buller 1983; Bright 1993; J. Marshall 1995; Harrod 2000); also by the treatment of corn by the Cherokee (*Tsalagi*), who consider corn to be a significant creative force in their tradition (Awiakta 1993).

This employment of nonhumans as creator figures for human cultural and spiritual traditions raises an important philosophical question: How does it change the worldview of a people if the entity that is given responsibility for creating their culture is not a human, or even humanlike? Additionally—not only is the creator nonhuman, but it is also a part of your current ecological community.

One certain consequence of viewing your creator as a nonhuman (animal or plant) is that it seems very unlikely that humans living under such cultural traditions would be troubled by the idea that humans, including themselves, came from organisms that would not be recognized as human. In Indigenous American traditions animal-persons and plant-persons existed before human-persons (Pierotti and Wildcat 1997a). This is an important point in the current debate over evolution, where one of the major issues is whether humans should be considered as having come from organisms that are not human, or alternatively if they were specially created in the image of an anthropomorphic creator figure, such as the Christian God. This also ties to the idea of relatedness. Keep in mind that in the clan systems of Indigenous Americans, totems are always assumed to be elders that existed before humans (Pierotti and Wildcat 1997a; Deloria 1999b).

If the entity you consider to be your creator represents a species of animal or plant that you are likely to encounter in your immediate environment and during your daily activities, this functions to maintain respect and affection for individuals of this species, as well as for the local natural world and its other inhabitants. At the same time this reinforces the idea of connectedness, by acknowledging a member of your ecological community as the originator of the local cultural tradition.

Viewing animals as creator figures associated with the development of new cultural traditions contains the clear implication that the animals existed before the humans did, since to function as a creator it is necessary to exist prior to your creation. Thus, it is clear that Indigenous Americans are aware that other nonhuman species existed in the world before humans did. In Rock Cree cosmogony, animals were recognized to have existed before human beings, and humans were known to come from animals during the regression of the earth (Brightman 1993). In the Lakota tradition it is recognized that "Sugmanitu Tanka Oyate (wolves), were a nation long before human beings realized and declared themselves a nation" (Manuel Iron Cloud, in McIntyre 1995).

An important question that arises is how such creator figures can be linked to the phenomenon of biological evolution as originally outlined by Charles Darwin in the middle of the nineteenth century, and as currently conceived by the modern scientific tradition (exemplified by Mayr 1961, 1997; Browne 2002; Gould 2002; Kirschner and Gerhart 2005; Carroll 2006). The concept that humans and nonhumans are related remains controversial to this day, particularly in U.S. culture and philosophical tradition (Miller et al. 2006; Petto and Godfrey 2007). Many major philosophers and scientists from the Western tradition place great emphasis on defining and justifying the separation of humans from the remainder of the natural world (Mayr 1997).

Employment of nonhumans as creator figures is conceptually related to but qualitatively different from the Western scientific concept of biological relatedness. In my opinion this difference arises for two primary reasons: (1) the Western scientific tradition has ignored the influences of nonhumans in shaping human cultural traditions and locally based ecological knowledge, and (2) Indigenous American traditions are based on understanding relationships as indicated through social behavior and ecology rather than through anatomical or genetic similarities (Barsh 2000). The Western scientific tradition has also failed to deal appropriately with the emotional components of interactions between humans and nonhumans as humans have co-evolved with other species (Morey 1986, 1994; Schleidt 1998; Pierotti and Wildcat 2000; Barsh 2000; Coren 2008). As discussed in previous chapters, emotional links based on relationships between humans and nonhumans are largely responsible for the development of spiritual traditions and ceremonies grounded in local ecological relationships and intended to reduce the negative impacts of environmental variation (Anderson 1996; Pierotti and Wildcat 2000, 2001).

GENETIC AND ECOLOGICAL RELATEDNESS

One basic premise of Indigenous thought across a wide range of cultural traditions is the relatedness of all life forms (Kidwell et al. 2001). I have addressed this above in the sections on Atleo, Deloria, and Tinker. The Lakota people of the northern American Plains institutionalize this belief by concluding all prayers with the invocation *Mitakuye oyasin*, which translates as "all my relations," signifying recognition of connections and relatedness to both the human and the nonhuman world (Nerburn 2009). Acknowledgment that human and nonhuman are related in a real and meaningful sense is a fundamental principle of evolutionary biology. In the Western scientific tradition, however, this relatedness is established through two factors not usually considered by Indigenous Americans: (1) recognition of the fact that DNA and RNA are the common hereditary material of all living creatures, and (2) anatomical similarities (homologous structures) between humans and various kinds of nonhumans such that the more similar the anatomical features and DNA sequences, the more closely related to humans are the nonhumans (see also Barsh 2000).

This recognition creates part of the difference between Western and Indigenous concepts of relatedness. In contemporary Western science, relatedness of humans to nonhumans is typically established through the study of similarities in sequences of nucleic acid/base pairings. Results of such investigations indicate that the closest living relatives of *Homo sapiens* are the anthropoid apes, including bonobos and chimpanzees (*Pan paniscus* and *troglodytes*), gorillas (*Gorilla gorilla*), and orangutans (*Pongo pygmaeus*). These genetic relationships are supported by anatomical evidence, such as the bone structure of the hand and skull and the appearance of the teeth.

Despite this unequivocal evidence of relatedness in a physical sense, Indigenous American traditions do not treat apes as close relatives. If humans were perceived by Indigenous cultural traditions as being closely related to anthropoid apes, it would seem that an appropriately designed creation story should employ such entities as creators. Nonetheless, no Indigenous peoples of whom I am aware regard apes, or other primates, as creator figures (with the possible exception of China where monkey is seen as a trickster), and this is unequivocally true of the Indigenous peoples of the Americas.

I believe the reason that large primates are not considered as creator figures by Indigenous Americans is because such creatures were not part of their direct ecological experience; hence they were never an important component of their cultural heritage (the story of their people). The only place that large primate figures appear in stories of Indigenous Americans is in the role of "primitive" manlike creatures who haunt the margins of human society, functioning either as bogeymen or as powerful and mysterious entities who should be avoided, such as the Woodsman of the Koyukon people in interior Alaska (Nelson 1983), Dzonoquah of Northwest coastal peoples, and Sasquatch of the northern Cascades.

The nonhumans attributed with creator status in Indigenous American cultures are frequently the most intelligent and social creatures of their particular ecosystem or community, and are regarded as the most powerful spirits of a particular local ecosystem whose existence provided both a livelihood for humans as well as a framework within which cultural traditions could develop. As discussed previously, the Western tradition of viewing human history as independent of place and ecology is not part of the conceptual framework of Indigenous peoples of North America (Deloria 1992;

Basso 1996). To Indigenous Americans, ecological knowledge and spiritual tradition cannot be readily separated. Thus, religion serves to code ecological knowledge, and this knowledge connects Indigenous peoples to the place from which they come (Rappaport 1971; Deloria 1992; Anderson 1996).

Understanding emotional (and spiritual) links is extremely important in terms of the way relationships among species are understood. Local wolves, deer, bison, birds, trees, herbs, and even rocks were considered to be closely related by Indigenous people because of their cultural relevance, combined with the fact that they were of the same place (Anderson 1996). Other humans, although similar in appearance, were considered to be more distantly related simply because they did not arise from the same place, hence they were not part of the local ecological community.

As an example, in the tradition of the Nuhmuhnuh (Shoshone and Comanche peoples) Wolf was considered to be the creator figure who created a perfect world (A. Smith 1993; Harney 1995: 26; Papanikolas 1995). The fact that the word *Numuna* (Nuhmuhnuh) translates literally as either gray wolf or human being (Ramsey 1977: 231) suggests that humans and wolves were considered to be equivalent as persons, which follows from their ecological equivalency within the Numic tradition.

This conception is almost certainly based on the close ecological relationships that existed between these two species as hunters. Humans and wolves share a long ecological relationship and probably a co-evolutionary history (Morey 1994; Schleidt 1998; Sablin and Khlopachev 2002; Coren 2008). Wolves are the direct ancestors of dogs, who are generally considered to be the first animal species to be domesticated (Clutton-Brock 1984, 1995; Morey 1994; Serpell 1995; Vila et al. 1997; Coren 2008). Humans and wolves, both wild and domestic, have probably shared cooperative foraging relationships for perhaps as long as 100,000 years (Morey 1994; Vila et al. 1997; Schleidt 1998), and the association with wolves may be a major reason that modern humans outlasted Neanderthals (Coren 2008). In addition, both species would at times feed from remains of kills made by the other or share kills (Schlesier 1987; J. Marshall 1995), and both species live in close family groups dominated by a mated pair of adults who train their offspring in survival skills.

The ecological, and especially the social, similarities between humans and wolves could be interpreted from a behavioral and ecological perspective as evidence of close relatedness. In the Western scientific tradition ecological similarity can lead to organisms being considered as members of the same taxonomic unit; the *ecological species concept* defines a species as the set of organisms inhabiting a single ecological niche (Ridley 2004: 353–354). The ecological niche is composed of the set of resources and habitats exploited by a species. Following this logic the tendency of wolves and humans to work closely both within and between species and to behave in a similar manner led to their being considered one another's closest relative by many Indigenous peoples of the Great Plains.

Wolves were the most visible social and intelligent predators in the environment of the Great Plains. In the Pacific Northwest, the most intelligent and social animal that is highly visible and interacts with humans on a regular basis is Raven. Ravens differ ecologically from wolves in that they are generalized omnivores and predators on small animals as well as scavengers on carcasses of salmon and large mammals; however, ravens are also monogamous and live in extended family groups that show complex networks of cooperation (Heinrich 1989, 2000). Raven is regarded as both a creator and trickster figure by tribes in the Pacific Northwest and Alaska (Nelson 1983).

The nature of hunting relationships and the prey exploited are very different on the American plains than in the dense woods and cold climate of the Pacific Northwest and Arctic. Ravens are not considered to be close relatives and equivalents of humans in the same way that wolves are. Wolves are considered to be creators in areas where humans and wolves exploit similar types of prey, such as bison, pronghorn antelope, and elk. Ravens are considered to be creators in areas where they help humans locate prey such as caribou (Brody 2000; Martin 1999; Cordova 2007: 12–13).

Thus, it appears that nonhumans regarded as creators are typically those species that are most intimately linked to the ecology and lifestyles of the human cultures in a particular ecosystem. In the northern forests where bears helped humans learn about plant foods, they were sometimes considered as creator figures (Rockwell 1991). These relationships illustrate the significance of the way in which the local human culture regards the role of these nonhumans in the establishment of their traditions that validate them as a distinct cultural entity.

THE NATURE OF CREATION

Connected to the idea of relatedness in Indigenous knowledge is the concept that each organism is an individual, with unique qualities. As discussed in the previous chapter, the statement made by Okute clearly reveals an understanding that individuals vary within species, and that the environmental conditions under which animals exist shape their appearance and behavior. Okute includes humans within his observations, which indicates that he is aware that humans evolve in the same manner that nonhumans do.

There are numerous points of convergence between Indigenous thought and evolutionary theory. Although Indigenous views are often described as "traditional," this should not be taken to mean that they are unchanging. Indigenous worldviews portray the universe as a system in continuous flux, driven by both known forces and powerful random elements, which are identified as "tricksters" (Barsh 2000). Since many aspects of the environment are bound to change in ways that cannot be forecast accurately, it is assumed that both humans and nonhumans can also change to respond to these changes, as described by Okute (above). One major feature of traditional beliefs is that they have existed long enough for long-range consequences to affect them.

All Indigenous traditions seem to readily accept the relatedness of all life forms, and some extend this concept of relatedness to the physical world (rocks, bodies of water, winds) as well (Cajete 1994, 2000; Kawagley 1995; Deloria 1999a, 1999b; Atleo 2004; Tinker 2004; Nerburn 2009). At the same time several of these scholars stumble over the issue of creation. The major exception seems to be Cajete, who states, "The idea of evolving or changing through generations, is part of Indigenous thinking," and "Everything has a time and an evolutionary path" (Cajete 2000: 36, 77). The difference seems to be that Cajete has scientific training and understands that evolution emerges from an understanding of relatedness and changing environmental conditions.

The confusion over creation among other Indigenous scholars appears to result from their having been raised as Christians and their inability to recognize the link between rejection of evolution and rejection of relatedness between humans and nonhumans. The Western monotheistic religious tradition posits a creator that is human in both form and thought. This creator, or god, is then assumed to have human limitations and human values. It seems that many followers of the Western philosophical tradition assume that if God does not think like them, then he cannot think at all, and therefore does not exist (Davies 1994: 77–78). This conundrum reveals the limits of the Western philosophical tradition and why fundamentalist Christians have resorted to the idea of intelligent design, which seems to assume that the creator functions as a master engineer (Shermer 2006; Petto and Godfrey 2007). In this tradition either God is the smartest human imaginable, or else he cannot exist.

In contrast, try to imagine a creator who is not human or even remotely humanlike in its thought patterns. This entity would not see humans as superior to or above other life forms because all life forms are its children. In this worldview, the world undergoes many changes. During these environmental changes some life forms may go extinct, while others survive even through extreme changes in environmental conditions. The life forms that persist, either as individuals or populations, are not "favored" or "chosen"; they are simply the organisms able to survive and reproduce in the changed environment. Those that do not persist must return to the earth, and their components will reappear as part of new forms of life. Thus, it is environmental change that helps to "shape" the life forms to come, so these changes can be described as drivers of the process of "creation." This way of viewing the world is entirely consistent with Indigenous stories and ceremonies that acknowledge the existence of a changeable world, in which those changes are often unpredictable.

One important consideration of this viewpoint is that humans are included among the life forms that suffer the consequences of changing environmental conditions. Therefore, as a consequence of these changes some human cultures may disappear, while others may persist. Indigenous cultures were aware that the world changed and that often when it did, some species, or even human cultures, were unable to persist or thrive. This serves to make Indigenous peoples "ecological" in a manner that is not often recognized, because they lived and died alongside other members of their ecological communities.

Many Indigenous cultures survived and a few have even thrived, but the impact of disease introduced by European invaders apparently led many Indigenous people to feel as if the world had turned against them (Martin 1978). This social and demographic devastation was probably enhanced by the apparent success of Europeans—they did not suffer as much from the diseases they had brought. This combination of factors may have led many Indigenous people to accept European religions and spiritual traditions, including the concept of a Big Man God who was obsessed with every detail of the lives of his followers.

Perhaps we should consider a different type of creator. With regard to the origin of life, and of the universe, we will never have unquestionable proof of what took place. Evidence can be gathered to support one perspective or another, but absolute proof will probably always be lacking, which is why the origin of life is not really an evolutionary question (Shermer 2006). The key point on the origin of life is that regardless of exactly how it happened, it took place billions of years ago. Since modern humans did not exist until the last few hundred thousand years, whatever the creative force was at the beginning of life, it certainly was not human. In consequence, I look to the one entity that I can be sure was in existence at that time, the Earth itself. To me Mother Earth, with all of its changeable faces and moods, serves as the most obvious creator imaginable.

Science does have one major advantage: It can gather and examine evidence, which leads to new ideas. All that fundamentalist religion and its stepchild, intelligent design, have to offer is faith; there is no evidence or experimental tradition that can be used to test their assertions (Petto and Godfrey 2007). Faith can be comforting, but a faith that requires the idea that the world has no meaning without humans and that the earth was created solely for the use of humans entails too high a cost, both at the ecological and spiritual levels.

In the Indigenous tradition evidence is also gathered and evaluated. Unusual occurrences are taken more seriously than in the Western tradition, but they are not considered to be "magic" or "miracles." Instead they are considered to be natural phenomena that occur in the natural world and that require careful evaluation. If an idea or concept appears no longer to work or be useful, new ideas can be developed to replace the old ideas as they have been developing for thousands of years. As far as the beginning of life is concerned, people can believe pretty much what they want; however, the earth is obviously real and essential for all life regardless of belief system or worldview. Without Mother Earth, life is impossible, and our Earth is the only place in the universe where we can be absolutely sure that life exists. Unless something dramatic and unprecedented takes place in the future, changeable old Mother Earth serves as a perfectly satisfying creative force in my view, and I can only hope that those who desire to believe in a different sort of creator would treat her with the respect they afford to their various "gods."

6 Applying Principles of TEK Within the Western Scientific Tradition

> "Doubt and wonder mean that the argument is not over the ownership of the past, but over the processes of the future informed by the changed and changing past."
>
> W. S. Penn (Nez Perce) (1995)

Most people would be surprised to hear this, but I think Charles Darwin's classic book, *On the Origin of Species by Means of Natural Selection* (1859), was probably the first Western work of scientific scholarship to show convergence with Indigenous concepts of how the world functioned. Many might select Thoreau's *Walden* as the exemplar of this tradition, but to me Thoreau is too confused, and in his own words, "Man cannot afford to look at Nature directly . . . He must look through her and beyond her" (McIntosh 1974). In contrast, Darwin stared directly at Nature, both at her beauty and her dark side, and emerged with a view of the world that changed the relationship of humans to Nature forever in the Western intellectual tradition (Worster 1994; Dennett 1995).

Darwin's great book has been described as a "catastrophic blow to human privilege vis-à-vis the species question" (Rohman 2009: 1), because, as stated previously, the Darwinian model changed the position of humans from separate from nature to part of nature and subject to the same "natural laws" as the rest of nature (Mayr 1997: 182). Darwin's work served as one of the "great ideological pivot points of the modern era," and his concept that "chance undergirds the natural order flew in the face of moral and religious concepts that . . . (bestowed) divine favor on (humans), who were set above the rest of the animate world" (Rohman 2009: 2). As a consequence, the issue of humanity's link to the rest of the animal world appeared to threaten the concept of the immortal soul and hence the traditional foundations of morality (Bowler 1990). Darwin himself described Nature as "a web of complex relationships" and argued that the economy of nature could be considered as a system of "places" (Worster 1994: 156).

Darwin himself might have been shocked to hear that his ideas converged on those of other humans that he considered primitive (Browne 2006; Ruse 2008), yet it is clear that his ideas established the basic truth that underlies "all things are related" and that humans are merely one species among many. More to the point, his great book, which the ecologist Paul Ehrlich has argued should have been called "The Differentiation of Populations by Means of Natural Selection," described how changing environmental conditions led to differences among local populations that resulted from differential reproduction among variable individuals. This is essentially the same phenomenon described by Okute when he stated that, "From my boyhood I have observed leaves, trees, and grass, and I have never found two alike. They may have a general likeness, but on examination I have found that they differ slightly. It is the same with animals . . . It is the same with human beings, there is some place which is best adapted to each . . . *An animal depends upon the natural conditions around it. If the buffalo were here today, I think they would be very different from the buffalo of the old days because all the natural conditions have changed*" (McLuhan 1971: 18–19).

One ironic aspect of the Darwinian revolution was that it took a long time to come to fruition. At the turn of the twentieth century it was largely displaced by the reductionist science of genetics, which argued that mutation, rather than selection, was solely responsible for evolutionary change (Browne 2006). During the late nineteenth and the first four decades of the twentieth century the developed world was obsessed with heredity and its ugly cousin, eugenics, ironically led by Darwin's charismatic cousin, Francis Galton (see Browne 2006 for a review of this topic). It was only when the eugenic movement was discredited because of its link with German National Socialism, that the neo-Darwinist movement was able to resurrect discussion of the relationship between selection and speciation (Mayr and Provine 1998; Browne 2006).

During the period when evolution by natural selection was being resurrected, a new field of science, ethology (a branch of animal behavior), began to attain prominence in Europe. Interestingly Darwin was also one of the pioneers in this field, with his publication of the books that followed the *Origin*: 1) *The Descent of Man and Selection in Relation to Sex* (1871) and 2) *The Expression of Emotions in Man and Animals* (1872). What is noteworthy is that even the titles make explicit the relationship between humans and nonhumans, an idea that was very controversial at the time. In the third chapter of *The Descent of Man*, "Comparison of the Mental Powers of Man and the Lower Animals" Darwin argues that nonhumans are probably self-conscious and that "there is no fundamental difference between man and higher mammals" (86), that both human body and mind are descended from nonhumans, and "difference between man and higher animals . . . certainly is one of degree and not of kind" (151) (Rohman 2009: 4–5).

At first, ethology adhered to neo-Darwinian orthodoxy, and focused on species typical behavioral traits, which were assumed to be evolved character states comparable to physical features. Things began to change in the early 1960's, however, when a young British woman, Jane Goodall, decided to devote her life to the study of the social behavior of wild chimpanzees (Goodall 1986, 1990; Peterson 2008). Goodall's way of thinking seems closely linked to the ideas laid out by Darwin in the preceding paragraph, and she changed the face of animal behavior for two reasons: 1) she was one of the first female field biologists and actually lived amongst her study subjects, which 2) allowed her to concentrate on the behavior of individual animals and their relationships to others within their social group. Goodall's careful observational techniques and familiarity with every animal in her study population revolutionized animal behavior (Peterson 2008).

Around the time that Goodall was initiating her life's work, the British ornithologist John Crook was conducting long-term field studies of weaverbirds (family Ploeciidae) in Africa (Crook 1964). These investigations, which focused on longterm studies of organisms living in their natural environment, began to shift the emphasis of biological research to field work, although very few investigators were willing to devote the time and energy to get to know a species population the way that Crook and Goodall did, which converged both in knowledge and concepts upon the TEK of Indigenous peoples.

FINDING NATURE IN SCIENCE

I began my university education in the mid-1960's, around the time that Crook and Goodall were publishing their first results, although it took several years for me to become aware of their work. In contrast to my early experiences with the natural world, my first experiences with academic science were unsatisfying. Even though I was drawn to nature, I often considered abandoning the study of Biology. The Biology course I took in high school had nothing to do with the fascinating world I had found around me as a young boy; it only seemed to involve cutting up other organisms to see how they were put together. Although I didn't know it at the time, this was the classic expression of the Cartesian machine metaphor and the tendency of the European intellectual tradition to emphasize parts rather than wholes. I felt that the natural world was part of my home and that it contained cultural meanings which family tradition had given my external surroundings. A statement made by a Lakota elder exemplified my feelings towards biology, "We aren't the collecting people, maybe we are the listening people . . . we can't take everything apart as well as you can . . . we can help you remember how it went together" (Nerburn 2009: 175).

When I went off to university my intention was to major in the arts, or perhaps the humanities, rather than in science. I was assigned a biology professor as an advisor, who suggested that I enroll in a new course the university was offering in a topic called Ecology, being taught by a visiting faculty member from Canada. In the mid-1960's ecology was not taught in most universities and in those where it was taught it was a recent addition to the curriculum. Rachel Carson's book *Silent Spring* had been published only a few years earlier (1962), and no university offered courses in Environmental or Ethnic studies of any kind.

I had not heard of ecology previously, therefore, I had no idea what this course involved. Once I learned what was involved in this course, I realized that there was still hope for me to have a career in science, pursuing this "new" discipline, because it involved the type of field related activities in which I had been participating since I was a young child. At its essence this course was about understanding connections both between different species and between biotic and abiotic, or physical, components of the environment. It was also the first science I had experienced that helped me to "remember how it went together."

I ended up attending several colleges in an attempt to balance family obligations with coursework while seeking out a friendly and supportive academic environment. My family moved again and family ties and obligations drew me to California, where I enrolled as a biology major, this time at the University of California at Santa Cruz.

UCSC was small (roughly 1800 students in 1968) and located on a former ranch that included old fields, oak savannah, and redwood forest. In 1968, the UCSC campus was infused with the natural world and ecology permeated the study of Biology. Lab courses frequently took students outside where we could visit natural communities on short walks or drives from campus. I finally found myself exposed to the studies of Jane Goodall and John Crook in another course that had not been offered in my previous schools, Animal Behavior, where the professors actually followed the lead of Goodall and Crook, studying individual animals interacting with each other and their natural surroundings.

As emphasized in previous chapters, Indigenous peoples characterize ecosystems as social systems, and in Indigenous knowledge, behavior and social interactions are the focus of inquiry, rather than details of anatomy or populations (Barsh 2000: 160). My initial frustrations with academic biology resulted from the tendency of Western science to focus on anatomy and cell biology as the basis of understanding adaptations (Barsh 2000; Lewontin 2001). Anatomical traits only undergo major change over a long period of time, which is one reason why Darwin and his followers assumed that evolution resulting from natural selection could not be directly observed on a short time scale.

Indigenous peoples focus on behavioral co-evolution between species (Barsh 2000), which can be observed as a local, place-based response that can happen over relatively short periods of time. In contrast to the evolution of anatomical traits, these behaviorally and ecologically driven traits result from the organisms' responses to local environmental changes and can happen almost instantaneously (Weiner 1995; Lewontin 2001; Kirschner and Gerhart 2005). The ability to respond quickly to environmental change lies at the basis of Indigenous philosophy. Statements cited above from Okute almost certainly refer to behavioral and ecological changes. Indigenous people participated directly in shaping the behavioral adaptation of other species, functioning either as predators or competitors of these organisms. In some cases, such as traditional burning, they even produced the environmental modifications that led to adaptive responses on the part of other species (Barsh 2000; Williams 2002).

Once exposed to both animal behavior and ecology I had a pretty good idea of what I wanted to study and how to incorporate what I perceived as Indigenous traditions into science. The paired ideas of connection and relatedness have subsequently informed my approach to science and the questions I ask when I examine the natural world. I was raised to view nonhumans as distinct beings that functioned within a strong social (population) and historical (evolutionary) context, and to pay close attention to individuals within populations in order that I might learn how they lived. Observers who interact over a long time with animals learn subtle cues to their behavior (Barsh 2000). Participating in such long-term interactions leads the observer to develop a relationship with and a sense of obligation to these other beings (Anderson 1996).

I began my academic field work as a volunteer assistant to professors who were studying Northern elephant seals, *Mirounga angustirostris*, and Steller sea lions on Año Nuevo Island about 20 miles north of the UCSC campus, during which we marked as many animals as possible so they could be individually recognized following in the traditions established by Goodall.

In December male elephant seals arrive from their feeding grounds in the Gulf of Alaska and compete with each other for potential access to the large number of females that will give birth on the beaches of the island. The basic metaphor employed in this research was that the males formed a "dominance hierarchy," a concept that assumes that animals have a hierarchical social system in which one individual, usually referred to as the "alpha" wins all conflicts and a similar "peck order" is followed throughout the society, with each individual "dominating" all individuals below it.

The male everyone assumed was on top, the so-called "alpha" male, named RAT (names were written on the sides of seals using Ultrablue hair dye), seemed mentally unbalanced. RAT was relatively small for an adult male elephant seal, which meant he probably only weighed a couple of tons, but he was highly aggressive, threatening, and chasing any other male who moved around him. As a result, most males tried to avoid catching his attention.

Male elephant seals display to one another by rearing up, lifting their head and neck clear of the ground and producing reverberating calls by dangling their proboscis (from which they get the name "elephant" seal) over their open mouths. Their mouth and throat serve as an echo chamber that adds to the resonance of the call. Males recognize other individuals by their calls, because they respond according to their relationship with the other male. If the calling male is dominant, the other male will either lie low or stop whatever activity he is engaged in, such as sneaking up on a potentially receptive female. A dominant male can also respond to the challenging male, in which case the latter will typically collapse like a balloon that has lost its air and lie low, hoping not to attract further attention. Only if the males are relatively close in "dominance rank" will both continue to challenge each other to the extent that actual physical combat (fights) might ensue. Fights involve the males rearing up, slamming into one another, and biting and tearing at the chest pad of the opponent with their two to three inch long canine teeth.

Watching the males interact it became obvious that the adult males (who are at least eight years or more old) distinguish between calls made by other adult males and the raspy, almost adenoidal, calls of immature males who have not developed a full proboscis (anywhere from three to seven years old). Adult males usually ignore the displays of immature males unless they are very close by, in which case they rise up and give a short perfunctory display that invariably serves to silence the immature.

RAT responded differently than any other adult male. He did not seem to discriminate between other adults and immature males, and quickly charged and attacked any male who displayed in his vicinity. The adult males usually were able to retreat, but when RAT caught immature males he mounted them like they were females, which they vaguely resembled. Male elephant seals are much larger than females of their kind, but this difference is not fully apparent until the males reach three to four years of age. It was possible that RAT didn't know the difference between sub-adult males and adult females, but that seemed unlikely. It felt more like he needed to humiliate any possible rival, especially those he could easily push around.

Female elephant seals become sexually receptive a month after giving birth. During that month they fatten up their pups with milk that's half fat and comes out of their nipples like toothpaste squeezed from a tube. The first babies are born around the winter solstice, with the peak of births occuring during the first two weeks in January. As a result, most females become receptive to males in early to mid February, at which time competition among the males heats up considerably. RAT expended a lot of energy chasing and attacking males before the females became receptive, which is the elephant seal equivalent of hitting lots of home runs in batting practice.

Near the peak of female receptivity in the second week in February two new large males appeared. One was a huge old wrinkled male, named ADR, who had been the top male in terms of females covered during the last two years. ADR was at least thirteen years old, possibly older (his birth occurred prior to initiation of the study). This year ADR had first showed up on the main beach in late December. RAT challenged him and he quickly retreated. Since that time he'd been biding his time on the other side of the island with a smaller group of females. Everyone associated with the project assumed ADR had seen his last days on top, so we wondered why he had bothered to return to the main beach. The next day the other new large male, Old Blue, challenged RAT, who responded as usual with an attack, even though he was worn down from chasing "rivals" during the previous month. RAT and Old Blue had a bloody battle that lasted close to an hour, with Old Blue winning. At the end, RAT retreated, his reign of terror over. We all assumed that Old Blue was the new "alpha," but the dust had barely settled from the battle when ADR reared up and charged Old Blue, who did not act at all like an alpha, but turned and retreated. ADR was on top again and remained the "alpha" for the remainder of the breeding season, covering more females than any other male for the third straight year; Old Blue was second, with RAT a distant third.

This result surprised other people working on the project, yet I found myself wondering if RAT had ever really been the alpha in the minds of his fellow elephant seals. He had conducted himself in a manner I associated with various bullies I had met during my life. He never seriously challenged or fought other males larger than him. What usually happened is that he would charge and other adult males would retreat, including ADR. We had assumed this meant that RAT was dominant and the other adult males yielded to his "superior rank." In retrospect, an equally valid interpretation was that when it wasn't worth fighting, other males had not wanted to fight an overly aggressive and possibly mentally unstable opponent. All these males knew each other because they had interacted over their entire lives. When the stakes became high enough, males who weren't intimidated by RAT's aggression decided to challenge him. Old Blue could defeat RAT, but ADR was the male to whom the other males deferred, just as they had for the two previous years. ADR took over the top spot without a challenge and without injury. He had waited until the moment was right and then seized the opportunity presented by the battle between RAT and Old Blue. Once ADR replaced RAT there was less chaos on the beach and survival of the young pups increased markedly.

Observing this dynamic, I learned a lesson from the nonhuman world about individual personalities and intelligences. ADR employed an obvious strategy, waiting patiently for the right moment (see also de Waal 2001, 2009). He never challenged RAT himself, but his size and age allowed him to assert himself once RAT was displaced. ADR was not highly aggressive towards other males. He almost never fought, his status was such that a challenge from him could stop fights between other males and displace any challenger. ADR tolerated Old Blue, who was his most likely challenger. As a consequence Old Blue also did well in the mating competition. A major fight between the two of them could have had harmful consequences to one or both which they avoided.

In contrast, for all of his aggression RAT was not successful. The next year he reappeared as a "low-ranking" male, and was caught by another large male while he was trying to mount a female. This male flung his body upon RAT while RAT tried to escape. In the melee RAT suffered a compound fracture of the baculum. He crawled to the water in obvious pain and was never seen again.

Watching the social interactions of these large slow-moving mammals, led me to realize that these were distinct beings, each one responded to specific environmental and social situations in unique ways. Reducing this to simple concepts like "dominance hierarchy" or "territorial defense" oversimplified the complex dynamics. The basic set of behaviors employed by each individual may have been "species typical," but how they employed them varied in relation to social relationships among the individual seals.

This was different than the ethological thinking that dominated animal behavior at that time. Ethology was developed by Niko Tinbergen, Konrad Lorenz, and other European scientists as a way of studying animals under natural conditions (e.g., Tinbergen 1953). Ethology was meant to replace Skinnerian behaviorism, which was completely mechanistic (Bekoff 2000). Ethology was still a mechanistic and reductionist approach to behavior, but at least it acknowledged the existence of the environment. Still, ethology was built upon the assumption that all individuals of a species would show the same behaviors, or "fixed action patterns," in response to the same stimuli (Tinbergen 1953).

Ethology was the dominant paradigm in Animal Behavior from the late 1940's until the mid-1970's, and it was the first paradigm under which I learned to study behavior. The intellectual heir to Ethology was Sociobiology, a theory derived from the idea that most social and cooperative behavior resulted from genetic relatedness between individuals (E. O. Wilson 1975). Sociobiology was just beginning to attain prominence as I began my career in field research. If anything, Sociobiology was more reductionist than Ethology in its fundamental assumption that animals functioned as genetically programmed machines, whose sole purpose was simply to struggle for reproductive advantage in natural selection (Bekoff 2000), or to "maximize their inclusive fitness" (E. O. Wilson 1975).

Sociobiological thinking heavily influenced the research group of which I was part, and I felt pressure to conform. I found some sociobiological ideas compelling, but remained troubled by the tendency to ignore, or at least avoid dealing with, my perception that the animals I watched were distinctive personalities. I watched intelligent calm tacticians like ADR, and spectacular failures who showed odd behavior, such as RAT. Both may have been "struggling for reproductive advantage," but it was obvious that individual style and personality, rather than genetically programmed tactics, determined the outcome. This realization gave me a principle that would inform my research for the rest of my career, that is, nonhuman societies have their geniuses and dunces, and even the occasional psychotic, just as do human societies.

The pinnipeds I watched spent most of their time lying around, resting from their exertions at sea. They are, after all, marine mammals and most of their important ecological activity takes place at sea where they are the masters of their environment, fast and graceful swimmers capable of diving more than a thousand feet in search of fish and squid. The following spring, while I was assigned to record the time and activity budgets of Steller sea lions, I began to spend my spare time watching another species that bred on Año Nuevo Island. This was the Western Gull, *Larus occidentalis*, a species better known to most people as a "seagull," even though there are several dozen species of gull around the world, some of which never visit salt water.

In contrast to the polygyny of the pinnipeds, gulls live in obvious family groups, with a male and female sharing a breeding territory on which they raise their offspring. I found the family dynamics of the gulls more intriguing than those of the pinnipeds. In the latter, a few males competed for access to a large number of females. This tendency of males to mate with multiple females fascinates male biologists, who I suspect often wish that they could behave in a similar fashion. One well-known animal behaviorist in the 1970's was known to tell women that in other species "females are attracted to high status males, *like me*." Some of my colleagues admired RAT's aggression and "machismo," and were upset when he came to his painful end, whereas I was not at all surprised by his fate. More important, in these polygynous species, males almost never interact with offspring, who in some cases may not even be theirs. To this day I remain fascinated by pinnipeds, but I also recognize that the large portion of their lives spent at sea will forever remain a mystery to humans.

I decided that more insight and understanding could be gained watching the lives of the monogamous gulls, which form pair-bonds that can last for as long as 20 years. Another rule I learned around this time was that gulls that weigh around two pounds and sea lions and elephant seals which weigh thousands of pounds had similar average lifespans (five to six years), and that many individual gulls outlive all of their giant mammalian neighbors. Wolves are one of the few monogamous species of mammal, but they are hard for a beginning scientist to study. My research career was oriented towards studying the behavior and ecology of monogamous vertebrates, so I chose gulls as a stand-in for wolves.

The gull population on Año Nuevo was small and easily disturbed, probably because they were not treated with respect and care by researchers studying marine mammals, and by the marine mammals themselves. I looked for a colony more remote than Ano Nuevo, where research on birds took precedence over research on marine mammals, and moved my research to the largest colony of Western Gulls in existence, on Southeast Farallon Island (SEFI). There I worked with biologists from Pt. Reyes Bird Observatory who were studying the population biology of the 12 species of seabird that bred on the island.

SEFI has become much more famous in recent years because of the great white shark (*Cacharodon cacharias*) population that forages around the island during the fall. The Farallon Islands have become reimagined as the

"Devil's Teeth," an appellation given them because they are a series of bare granitic peaks that poke out of the ocean some 25 miles west of San Francisco's Golden Gate (Casey 2005). In her bestselling book about the white shark population Susan Casey provides the following description of SEFI: "The islands themselves are fragile . . . made of 89 million-year-old granite, much of which has gone rotten and crumbles to the touch . . . The boat ride from the mainland is a riot of turbulence and nausea . . . The place gets regular lashings of the meanest weather the Pacific can dish out. Thirty-knot winds, blanketing fog, and 15-foot seas are standard" (Casey 2005: 6–7). Later on Casey informs us that the Miwok Indians considered them, "The Islands of the Dead . . . An island in the bitter salt sea . . . naked, barren, and desolate, covered only with brine-spattered stones and with glistening salt . . . and swept with cursed winds" (77).

I suspect that the Miwok considered SEFI to be "The Islands of the Dead" mostly because they are readily visible from the higher vantages of San Francisco and Marin, and they were located to the west, traditionally the direction associated with death in Indigenous traditions. Despite the wind, fog, and isolation I always found SEFI to be a comfortable, pleasant place to do research. After all, it had electricity, running water, a stove, and decent beds. I barely recognize the place as described in Casey's book and in televised specials on the shark research. To me and the 25,000 Western Gulls that call it home, SEFI was a very nice place to be, appropriately wild and surrounded by a productive ocean. Casey's book showed me again how alienated individuals of European heritage are from nature and how they feel the need to find nature fearful and brutal.

INDIVIDUAL VARIATION

One of the most interesting models that emerged from the Sociobiological tradition was the idea that patterns of parental investment (time and energy devoted to caring for offspring) determined whether or not males would form a pair bond with a female or would instead abandon a female after mating and seek other females (Trivers 1972). Sociobiologists assumed that these patterns would be species typical, and that there would be little important variation among individuals. The first research project I designed and conducted on my own was an attempt to evaluate this model by looking at patterns of individual male and female parental care in Western Gulls on SEFI. Most tests of parental investment theory at that time (the mid-1970's) emphasized "typical" roles of males and females and came to general conclusions, that is, that most mammals, where females carry out most or all offspring care, because of the physiological constraints associated with gestation and lactation, are polygynous, whereas most birds, where females lay eggs, and males can participate in both egg tending and chick feeding, are monogamous and form pair bonds between males and females (Lack 1968).

There are exceptions; for example, wolves, coyotes, and other members of the Family Canidae have extensive male participation in rearing offspring, which was probably one of the many factors that made them an appropriate role model for Indigenous peoples and other humans. In a similar fashion, a few groups of birds, such as pheasants, grouse, turkeys and their relatives, males mate with multiple females and show no parental care.

The most interesting thing about parental investment theory was that it could be applied to individuals, although no one else had done this. Based on my experience with elephant seals, I wanted to study how individuals varied within a species to see whether this theory was capable of more subtle predictions. Gulls were ideal for such a study because they nested in large aggregations and showed extensive care by both males and females.

The male and female gull in a pair constantly adjust to one another, so the behavior of individual pairs varies according to the foraging abilities of each parent (Pierotti 1981, 1987). If one female takes more time than others in finding food, her mate adjusts by spending more time sitting on the nest incubating the eggs than did other males. One pair featured a male with only one leg who was perfectly capable of defending a territory and catching fish to feed his mate, but he had one problem. Male gulls stand on the backs of their mates and lower their cloaca to contact that of the female in order to complete copulation. A male with only one leg is incapable of lowering himself into proper position. The mate of the one-legged male would wait until he left to find food and then she would solicit copulations from males on neighboring territories (Pierotti et al. 1996). This was the only female gull I ever observed to allow a male other than her mate to copulate with her. During more than a thousand hours of observation I noted only two situations in which female gulls would actually fight: 1) if their chicks were threatened, or 2) if a male other than their mate attempted to copulate with them.

Some general patterns did emerge, which represent the sorts of things that led ethologists to assume that species typical behavior existed, but there was a lot of variation among individuals in how these patterns were manifested. For example, females were strongly tied to the nest and the chicks, whereas males were more concerned with defending the territory (Pierotti 1981).

My first two years on SEFI studying gulls revealed considerable variation in the available food supply, which forced male gulls to adjust their parental activities to compensate for the extended foraging absences of their smaller mates, who had more trouble obtaining food under poor food conditions. In 1974 feeding conditions were better and females spent less time foraging, which allowed them to spend more time tending to eggs and chicks.

This variation in male and female parental care reinforced two things: 1) the flexibility of individual gulls and pairs, and 2) the importance of environmental variation. Gulls could not anticipate from one year to the next what weather and foraging conditions would be like. Observation of such phenomena led me away from the "balance of nature" and "constant environmental conditions" type of thinking I had learned in my first ecology courses, and I realized that all free-living animals and plants experience constant unpredictable fluctuations in weather and food availability.

Ironically, when I reported these results at a national meeting, a well-known ornithologist told me that, I was clearly wrong, because "evolution is over, it happened in the past." I found that this type of thinking appears to set limits that exist in the Western mindset. Like Okute, I expected the environment to change and the organisms to respond to these changes. What I saw from many of my Euro-American and European colleagues was the idea that evolution had happened in the past and that organisms were programmed to respond to an environment, which the scientists assumed to remain relatively constant. Given this type of thinking, it was hardly surprising that they expected every member of a species to behave in the same way.

The first year of my study (1973) was an El Niño year during which foraging conditions were quite poor, because the warm water associated with an El Niño reduces nutrient upwelling upon which plankton and fish depend, so the effect is observed throughout the system. The 1973 El Niño was one of the first El Niño /Southern Oscillation (ENSO) events to receive serious attention from ecologists, because it demonstrated clearly that environmental conditions did not remain constant. During the 1973 El Niño, foraging conditions were worse for female gulls than for males. To compensate for these time demands on their mates, most male gulls spent more time sitting on eggs than they did in 1974, when foraging conditions were better. Females also adjusted to situations that demanded the attentions of their mates. If the male of a pair was on the nest when an intruding male approached the territory, the female would rush to the nest and relieve her mate of incubation duties so that the male could engage the intruder (Pierotti and Annett 1994). Individuals in pairs that succeeded at reproduction adjusted constantly to each other's needs and preferences. If a male was late in relieving his mate, or did not sit on the eggs when she was gone, this could result in loss of eggs or chicks. Such losses were the primary factor that could lead to the break up of the pair.

These overall patterns were interesting but I also received instruction from the gulls themselves concerning how the details of their lives converged on humans. Some gull chicks starve because some parents aren't good enough at finding food and bringing it back, especially during El Niño conditions. Each year approximately half the gull chicks I studied died before they were old enough to take care of themselves.

Casey describes gull foraging behavior in emotionally loaded terms: "(Gulls) pillaged the murres . . . plus the cormorants and the auklets and any other birds that came around, and they killed their own too, with cannibal gusto" (2005: 107). I had a very different impression, backed by data. A few individual gulls specialized in feeding upon auklets, but they treated their own chicks with considerable care. Gulls never take adult murres or cormorants, and only rarely can they take chicks or eggs from these species unless humans cause a disturbance on the colonies.

Two particular experiences involving chicks changed the way I looked at gulls and their emotional responses. The first happened after I rescued a chick that had gotten separated from its parents. It was wandering about lost trying to find shelter or its nest. Other gulls were attacking it because they knew it wasn't their chick. I kept it in my room, fed it, and took care of it. The chick quickly became used to me and would run around my floor and fall asleep in my lap. One day the weather was unusually hot so I opened the doors and windows. The chick could hear the gulls on the colony and tried to get out. It struggled against the screen door that faced the colony. I wouldn't let it outside because I knew it would almost certainly be killed, but it continued to struggle to get out. Later that day I found it lying next to the door. I picked it up and it died in my hands. It was perfectly healthy and well fed. It just wanted to be a gull and be out with the others. When it couldn't achieve that goal it chose to die rather than accept captivity.

The second event occurred while I was watching a gull pair that had a nest right near my blind. Their nest was exposed and hard to defend, and this pair kept losing eggs and chicks. Finally they had only one chick left. One day while the male was absent finding food a gull swooped in and took the last chick. The female chased the other gull but there was nothing she could do. When the male returned with food, he gave the "mew call" which serves to attract both females and offspring to come and get food. The mew call is the most plaintive sound a gull can make and indicates a high state of arousal. When no one came to his call the male stood in the empty nest, his beak full of fish, continually making these sad-sounding cries.

These observations made it obvious that Darwin was onto something when he linked the expression of emotions in man to analogous behavior in animals. Despite the caricatures promulgated by journalists like Casey, gulls cared about and were devoted to their mates and offspring. Some were good parents and some were not. As I argued in the previous chapter, nature isn't cruel, but she doesn't play favorites. Offspring survive only if they have good parents who honor their responsibilities and take proper care of those offspring who are entrusted to them.

These experiences led me to recognize that chicks were also individuals and actors within the society. The prevailing Sociobiological theory viewed offspring as passive recipients of "parental investment," which depending on its quality could determine whether the offspring would live or die. Offspring were viewed solely as vehicles through which their parents could achieve reproductive fitness (Wilson 1975). Little was known about gull chick behavior other than they would peck at the red spots on their parents' beaks in order to stimulate their parents to feed them (Tinbergen 1953). Although all the other studies on Southeast Farallon involved monitoring the survival of offspring, I was the only investigator interested in the behavior and ecology of individuals, a situation that apparently continues to this day. Casey (2005: 132) states that "the biologists (on SEFI) grappled with the notion of animals as individuals." The other studies being conducted on SEFI dealt with collecting data on large numbers of birds and averaging these data in an attempt to characterize the population.

As part of my SEFI duties I was assigned to check several hundred gulls nests once every three days in order to obtain population data. This involved marking each egg as it was laid and each chick as it hatched. During collection of these data I found that after chicks hatched, a few chicks always seemed to be at nests other than the nest where they had hatched. Naïve and influenced by Sociobiological theory, I assumed that the chicks had "gotten lost" and would prefer to be back with their siblings and parents because they were close genetic relatives and theory predicted that relatedness trumped all other considerations in social interactions (E. O. Wilson 1975, 1978). If I returned these chicks to their natal nests, however, the next time I visited the study area I would again find the chicks somewhere other than their natal territory.

Initially, I found this frustrating. It was a good thing that the other two out of every three days when I was not checking nests I spent watching the family behavior of gulls, which revealed the answer to my confusion. It was obvious that considerable differences existed in the abilities of parent gulls, which could be seen in how often they fed their chicks and how vigilant they were at keeping other birds off their territories. If I could recognize this, the chicks probably could as well, because their lives depended on such information, whereas to me this was only part of a Masters' thesis. Chicks who had inept parents would abandon those parents and try to insert themselves into groups of chicks that had good parents.

The prevailing view of parent-offspring interactions at the time was a model called Parent-offspring conflict (Trivers 1974). The problem with the Parent-offspring conflict model is that, like other models in the Western tradition, it assumed that all parents and all offspring were alike, and that if an offspring was receiving less care than it wanted it would act out, which could force its parent to provide more care. If such conflicts arise in the real world, the parent always wins them. There are two reasons for this: 1) the parents are larger and more experienced, and 2) if the parent's life is put at risk by providing more care the parent will simply abandon the offspring (Pierotti 1989, 1991b).

As I wrote papers based on my research I read Leslie Silko's *Storyteller*, which includes a story of a mother who has to flee to high ground because of a flood while her daughters are away. When the children return they find their mother gone and the older sister comforts her younger sister by singing her the following song: "Little sister go to sleep . . . I suppose our mother didn't think much of us and she left us behind. By luck we might catch up to our mother who has gone ahead" (Silko 1981: 41).

These words brought to mind some of the gull chicks I had watched. Some parents simply were not able to provide more care. Acting out on the part of chicks would be pointless; they needed to find a way to survive, and used specific tactics when they made these changes. They sought nests where they were bigger than the resident chicks, because resident chicks did not want more siblings and resisted attempts to join them. Wandering chicks took advantage of disturbances on the colony to change territories. If adults were flying overhead they were less likely to see chicks entering their territories. Once on a new territory, chicks headed for the nest and crouched in the nest cup, because adult gulls will not attack a chick in the nest. A typical gull pair produces three eggs during a single season, which meant that no pair should have more than three chicks. Some gull pairs I watched ended up taking care of four or five chicks, whereas less-skilled parents had only one or two, in some cases none at all (Pierotti and Murphy 1987, Pierotti 1989, 1991b).

Such behavior did not work perfectly because adult gulls were not always altruistic. Some wandering gull chicks were attacked and even killed when they moved off their natal territory (Pierotti 1991b). Other investigators had noticed such dead chicks and tried to explain these deaths as misplaced "territorial aggression" by adults (Hunt and Hunt 1976). Such behavior was behind Casey's comment concerning "cannibalistic gusto." In fact an entire subfield developed within Sociobiology to explain the evolutionary basis of so-called "infanticide" in a wide variety of species (Hausfater and Hrdy 1984). In contrast to the accepted mechanistic explanations, I realized that to chicks who risked death while seeking better parental care this meant a quick death, rather than a slow one from starvation. Chicks were willing to take these risks because occasionally they were successful. Some chicks managed to get "adopted" by adults other than their parents, and had much greater chances of surviving (Pierotti 1989, 1991b).

There were scientifically accepted views of "alloparental care"—the term for care by an adult other than a genetic parent (Riedman 1988). The ethological tradition viewed this phenomenon as a result of errors on the part of "adopting" adults. The sociobiological tradition contended that alloparental care probably resulted from kin selection, and assumed that the caregiving individuals were genetically related to the offspring, which meant that they would be increasing their own genetic fitness through this behavior even if they were not the parents (Riedman 1988; Pierotti 1991b).

The prevailing explanations for both infanticide and for alloparental care proved unsatisfactory in accounting for the behavior I had observed. It seemed obvious that the chicks were driving the interaction, and that the adults were reacting in a complex fashion that depended upon both individual ability and ecological conditions. Good parents, who were able to collect lots of high quality food and worked well together as a pair, were more likely to adopt. More adults were willing to adopt when feeding conditions were good than when they were poor. This meant that the outcome of the interaction between chicks and adults could vary between years, even for the same individuals. The idea of kin selection did not allow for such variable outcomes, because it represents a mechanistic approach to explaining behavior that can be better explained in terms of social relationships.

The data I collected showed how individual adults varied in their ability to obtain food and bring it back to their mates and offspring. Gulls very conveniently regurgitate all of the food they have in their crop when they feed their mate or chicks. With a telescope or a good pair on binoculars it is easy to identify the food items, provided they are not overly digested (even then it is possible to tell if it is fish, squid, shrimp, etc.). Males and females took very different diets. Males take mostly large fish and squid, whereas females take mostly shrimp, crabs, and small fish (Pierotti 1981, Pierotti and Annett 1990, Annett and Pierotti 1989). Despite Casey's description, birds made up only a small fraction of gull diets. What was more important is that some males were much more efficient at finding food and bringing it back than were other males. Efficient males ended up feeding their mates and own chicks well, and were the ones most likely to adopt chicks that were not their own.

MICROEVOLUTION AS A PLACE-BASED PHENOMENON

While I was conducting my master's research on SEFI, two classic field studies were being initiated in other parts of the world. The results of these studies would demonstrate the underlying truth of basic evolutionary principles developed by Darwin, and accomplish this by employing techniques that converge on those employed by Indigenous peoples in establishing their own scientific knowledge about nature.

First, Peter and Rosemary Grant began their long-term study of finches on the Galapagos Islands. What made the Grants' approach unique was that it concentrated on local populations and knowing as many individuals as possible in a specific population and environment. The Grant's established one part of their work on the small island of Daphne Major, where they could band and recognize every bird in the populations of three different species. This work would demonstrate convincingly that the action of natural selection as described by Darwin in the *Origin of Species* could be seen over very short periods of time and that the strength of selection was related to the magnitude of environmental change (Grant and Boag 1980; Grant 1986; Grant and Grant 1989a, 1989b).

John Endler initiated the second important study, on guppies (*Poecilia reticulata*) in Trinidad. Endler looked at streams flowing down mountainsides and formed a series of pools that functioned as small and distinct habitats. Endler was able to demonstrate a balance between natural and sexual selection that depended upon variable ecological situations between different pools in the same stream (see Weiner 1995 for a good popular account). Endler's results showed the highly local nature of selection pressures upon specific populations, that is, within individual pools in a stream, a result emphasized both by Okute and by Paul Ehrlich in his studies on butterflies in California.

While these studies were getting started, I began a PhD program at a Canadian university where I would study two other gull species, the Herring Gull, *Larus argentatus*, and the Great Black-backed Gull, *L. marinus*, and their impact on breeding success in Atlantic Puffins, *Fratercula arctica*, on an island off the coast of Newfoundland. This allowed me to continue observations of parent-chick interactions, while I continued to collect as much data as I could on gull diets.

In Newfoundland there was a range of potential food types that varied in availability and abundance even within a single breeding season. Early in the season gulls could feed in the intertidal on mussels, crabs, and sea urchins, or they could travel to the local refuse dumps and take edible waste discarded by humans. A few birds even specialized on other smaller species of seabird that nested on the island (Pierotti and Annett 1990, 1991). In late May capelin, *Mallotus villosus*, a small energy-rich type of fish, came in to spawn in the inshore waters and along the beaches and most birds switched their diets to this rich source of food.

There was a surprise waiting. Gulls were fully capable of taking capelin spawning along beaches, however, most capelin moved about in large schools in nearshore waters. The gulls had found another way of getting schooling capelin. Gulls took fish driven to the surface as they attempted to escape from feeding humpback whales, *Megaptera novaeangliae*. Once I learned to watch for this interaction it was obvious that this was a major way for gulls to take capelin, even though every other biologist that studied gull foraging had missed this association. A key to this insight was that gulls did not start taking capelin until mid-June. This population of humpbacks breeds around the Dominican Republic in winter and in March they begin to migrate back to their feeding grounds in the colder more productive waters of the North Atlantic. Although capelin begin to arrive in nearshore waters in May, humpback whales don't arrive until mid-June.

My assumptions about foraging associations between gulls and whales were influenced by the Indigenous-themed idea that mixed-species groups of animals forage cooperatively, rather than competitively. The traditional relationship between foraging marine mammals and gulls was assumed to be competition, or at best commensalism, in the Western scientific literature. I saw that the association between gulls and whales is functionally similar to the relationship between coyotes and badgers, in which the cooperative foraging between these two prairie carnivores underlay why Indigenous peoples described these species as "friends" (Minta et al. 1992; Chapter 3). In my study, I found that large whales use the conspicuous gulls to locate rich concentrations of patchily distributed fish and squid. The underwater pursuit activities of these marine mammals force the prey to concentrate at the surface, where the prey is much more accessible to the gulls. As with badgers and coyotes, both groups have a higher rate of feeding success as a result of their cooperation (Pierotti 1988 a, 1988b, Pierotti and Wildcat 2000).

Another area of interest was how the choice of diet varied in relation to nesting habitat. A new branch of Evolutionary Ecology, called Optimal Foraging Theory (OFT) had been developed during the late 1970's. OFT is derived from economic models that were designed to optimize profit, and assumed that foraging organisms would always take the food type that yielded the highest rate of caloric intake (Stephens and Krebs 1986). Instead we found that food choice varied in relation to local area and that gulls did not take the food type that was highest in calories, but the one that allowed them to produce the healthiest offspring (Pierotti and Annett 1987, 1990, 1991). As discussed above we also found switches in diet that related to the hatching of eggs and the presence of small chicks (Pierotti and Annett 1987, Annett and Pierotti 1989). This suggested that gulls had a very different view of what an optimal diet was than did ecologists, animal behaviorists, or economists of European ancestry.

As on Southeast Farallon, adult gulls in Newfoundland varied in their ability to acquire food and care for their offspring (Pierotti and Annett 1990, 1991). Like their Western Gull counterparts, chicks of Herring and Great Black-backed Gulls tried to find ways to survive if their parents were not very good at parenting. The chicks of these species came up with a new wrinkle that I had not seen in Western Gulls; they formed groups or crèches. I first noticed this in Herring Gulls. In one area of the island I regularly found chicks from two or three broods hiding together when I ran nest checks. My previous experience convinced me that I should let the chicks make their own decisions. When I watched these chicks I realized they moved around as a group. When any of their parents returned, all of the chicks would run over and beg for food. The parents did not seem able to distinguish among the chicks, but it was clear that if it became necessary, the chicks knew both who their parents were and who the adults were who brought back the best food on a consistent basis.

Great Black-backed Gulls were more complex behaviorally. These large predators, who sometimes feed on other birds and their chicks, nested either alone or in loose groups of three or four pairs. Where they nested in groups, chicks from all the broods wandered around together, but unlike in Herring and Western Gulls their parents were completely tolerant of each other. They did not defend individual breeding territories. Adult Black-backs could land anywhere in the group area and feed chicks from any brood (Pierotti and Murphy 1987). Such apparent cooperation among large gulls had never been described. This behavior reminded me of the activities of a pack of wolves I had watched at my university's field research station, in which all the adults cared for the offspring.

Observing that parent-offspring interactions in Herring and Great Black-backed Gulls were similar to those in Western Gulls increased my confidence in my results. This was a general phenomenon that could not be easily explained by prevailing scientific thought. In all cases it was apparent that the chicks drove the dynamics of the system and the adults only reacted to a situation created by the chicks. I presented these results at a couple of national meetings and was told by a prominent sociobiologist, ironically the same individual who told Bernd Heinrich that ravens were probably fearful of predators (Chapter 3), that my results could not be right. He argued that any chicks that took advantage of adults would have the same thing happen to them when they became parents, and that such a system could never work. I pointed out that all chicks did not do this, only the ones who received inadequate care, but from his genetically deterministic point of view there had to be a "gene for such behavior" and such a gene "could never be favored by natural selection" (see also the quote from E. O. Wilson in Chapter 4)

This argument had one useful outcome. Developing a response to his argument revealed that the key to the interaction was the consequences were dramatically different for parents and offspring. An adult that took care of an unrelated chick expended some extra time and energy, but its life was not at risk. In contrast, to the chicks this was a life or death situation. If they failed to obtain adequate care they would die (Pierotti 1991b). By thinking of this behavior as a flexible phenomenon I was able to show that this conflict was a complex interaction between juveniles and adults. This flew in the face of the prevailing sociobiological view because it was obvious that chicks did not care if the adults that cared for them were genetically related, only that they were competent parents who provided adequate levels of food and protection. After all it doesn't matter if the adult who feeds and protects you is your genetic parent or not, as long as they do a good job.

I had little trouble writing my PhD dissertation. The trouble came in getting it accepted. There was resistance to some of my results, especially those concerning interactions between gulls and puffins. My emphasis on connections outside of the Herring Gull-puffin dyad, and the influence of Herring Gulls on the system went against established theory. I was able to show that gulls virtually ignored puffins in years when capelin were more abundant. My critics assumed that capelin were evenly distributed and consistently abundant from year to year (Nettleship 1972), an interpretation that was clearly wrong, but was firmly rooted in the idea of a balanced and stable nature to which some members of my committee adhered. A puffin biologist told me that "I would never get my gull-puffin results published because he was certain to be asked to review any manuscript on this topic" that I might submit. I was able to publish all of the important results from my PhD work (Pierotti 1982, 1983, 1987, Pierotti and Murphy 1987; Pierotti and Annett 1987, 1990).

While writing my dissertation, I learned about the important results coming from the studies being conducted by Endler and by the Grants and their students. It was gratifying to see that these results also deviated from the "balance of nature" thinking that had dominated early ecology. In particular the Grants results showed that an extended drought in the Galapagos caused significant changes in their study populations, and that only the largest birds with the heaviest beaks were surviving (Grant and Grant 1989a, 1989b; Weiner 1995). This fit closely with the Indigenous ideas that changes in the environment forced organisms to respond in ways that also left them forever changed.

After finishing my PhD, I studied Yellow-billed Magpies, *Pica nuttalli*, in the Carmel Valley in California. I applied the same approach to this study that I had to my gull work. I was pleased to have this opportunity because when I had been a young boy in Central California, Yellow-billed Magpies had been my most interactive and conspicuous nonhuman neighbors. One day when I was watching magpies feeding in an oak-savannah habitat an incredible ruckus broke out. A Cooper's Hawk, *Accipiter cooperi*, had attacked and killed a female magpie. Magpies from all over the area gathered in the trees around the kill site and chattered constantly while the hawk ate the magpie. This was not that surprising, animals often gather and observe a predator after it has taken a member of their group, however I did not expect what happened next. After the hawk left, the magpies flew down and walked over to the remains of the dead female. They no longer chattered, instead they muttered in low voices, like they were talking to each other. To my surprise, some magpies picked up feathers from the dead bird, took them into the trees, and stuck them there. After 15 to 20 minutes all the magpies except one flew away silently.

The only remaining bird was the mate of the dead bird. He picked up one of her primary wing feathers and carried it around with him for several days. When he stopped to eat he would put the feather down carefully and eat, then he would pick the feather up again and fly off with it. I realized that I was witnessing something akin to a funeral, or at least a celebration of a death, in a nonhuman. Other observers had described this behavior (Miller and Brigham 1988; Trost 1999), but they had not seen the feather carrying aspect.

I had been moved by what I interpreted as grieving in a male gull that had lost his last chick (see above). Being allowed to witness this magpie "ceremony" showed me what my mother and grandmother had meant when they referred to nonhumans as "people" and told me that nonhumans could engage in ceremonies of their own. It was obvious that this was a "ceremony"—every investigator who has observed this behavior has described it in these terms. Charles Trost, one of the leading students of magpies in the world took me aside at a national meeting and asked if I had ever seen a magpie "funeral." In his species account of Black-billed Magpies, Trost describes this behavior as "poorly understood" (Trost 1999:

11), which I take to be code for "it does not fit readily within the Western machine-based understanding of bird behavior."

CHANGES IN THE LAND

In the twenty first century, it is more permissible to speak of emotions and even of passions in animals, but this is a very recent development, only a few decades old. Jane Goodall was one of the pioneers of this field, but another excellent and insightful student of animal behavior, Marc Bekoff of the University of Colorado, pioneered the field of what he calls "cognitive ethology" (Bekoff 2000, 2002, 2006, 2007). Bekoff's work is particularly important for two reasons: 1) he has emphasized the importance of recognizing mind and emotion in nonhumans, and 2) he explicitly links such insights to the intellectual and scientific traditions of Indigenous peoples. Goodall and Bekoff have even worked together to produce books that emphasize these points (e.g., Goodall and Bekoff 2003). The excellent observer Franz De Waal has established that animals know right from wrong, and have political dynamics, and learn by carefully observing one another (de Waal 1995, 2001, 2009; de Waal and Tyack 2003). Excellent examples and discussions of this social complexity, emotional behavior, and cooperation among nonhumans can be found in those works by Bekoff and de Waal, as well as in the collection edited by Mitchell et al. (1997).

What is troubling is how long it took for this topic to reemerge from the NeoDarwinian closet, and how fearfully most scientists tread when they approach it. I assumed from the beginning of my research career that I was honored by being allowed to witness complex beings who responded to the world in many ways the same way that I did. As I progressed further into science I found myself under increased pressure not to address such topics, and I have been criticized when I suggested that cooperation as well as competition was important in understanding the ways in which species and nonhuman societies interacted.

I have watched fads like Sociobiology, which had some appeal to me as a beginning scientist, and extreme adaptationism, fall in and out of favor in science. When I first learned of Sociobiology (E. O. Wilson 1975: 1978) I assumed that, like Darwinian Natural Selection, it was a way of showing that humans were subject to the same biological rules and constraints as nonhumans. This turned out to be true, but it was also a view of nature that was opposite to Indigenous thought because it emphasized mechanical and deterministic approaches to understanding behavior. Sociobiological thinking seemed incapable of envisioning relationships and connections other than genetic ones (cf. E. O. Wilson 1998).

I have seen what are at best spectacular errors in logic, such as the idea that ravens were afraid of wolves rather than connected to them through relationships. Many scientists refuse to acknowledge that they are dealing with other complex beings with complex motivations. In Chapter 4, I discussed the obsession in Western science with identifying criteria that make humans unique from the rest of creation, even when it hurts science as a whole, and evolution in particular.

Over the 30 odd years of my research career I have become known for four principal discoveries: 1) the relationship between parental investment and mating in individual populations and social groups (Pierotti 1981, 1982, 1987), 2) The dynamic interaction between adoption and infanticide that led to the idea of the "intergenerational conflict" (Pierotti and Murphy 1987, Pierotti 1989, 1991b), 3) The study of dietary variation and its relationship to reproductive performance both on an annual and lifetime basis (Pierotti 1982, Pierotti and Annett 1987, 1990, 1991; Annett and Pierotti 1999), and finally 4) for showing a strong relationship between hybridization and male parental investment (Pierotti and Annett 1993; Good et al. 2000). The last two of these ideas were developed through collaboration with my wife, Dr. Cynthia Annett, who in both cases came up with the original ideas and showed me how these ideas and their applications fit within the Indigenous way of knowing.

All of these discoveries have emerged from emphasizing that nonhumans are complex beings, each capable of responding to a range of social and ecological situations. When I began my research career, the prevailing scientific theory was not much of a conceptual advance over earlier mechanistic models, and suggested that all members of a species should respond in pretty much the same way, which was clearly untrue. The nonhumans I studied all showed distinct individual personalities.

Each discovery has initially been controversial and encountered opposition, yet each has stood the test of time. I am encouraged by the fact that many contemporary scientists are moving more towards the study of individuals. In recent years the famous evolutionary biologist Richard Lewontin has emphasized the importance of the immediate environment in shaping evolutionary change by altering individual development (Lewontin 2001). Similarly the field of Evo/Devo (linking evolution with developmental biology) has shown that genes are less important and the environment more important in determining individual phenotypes (Gerhart and Kirschner 1997; West Eberhard 2003; Kirschner and Gerhart 2005). In many ways European and Euro-American scientists are rediscovering what Indigenous people have known for hundreds if not thousands of years (see Chapter 11).

7 Connected to the Land

Nature and Spirit in Native American Novels

What it boils down to is respecting your world, every little piece of it.
Hoey McCurtain, in Louis Owens, *The Sharpest Sight*

It might seem odd to include a chapter on literature in a book that deals mostly with natural science; however, in this case we are discussing a form of scientific knowledge that has been denigrated and ignored since it was first encountered. "The Indian explanation is always cast aside as superstition . . . reducing them . . . to a prehuman level of ignorance" (Deloria 1995). "Non-aboriginal people have been taught to despise, resent, fear or (at best) pity Aboriginal people, who will grasp any opportunity to 'get even'" (Barsh 2000: 154). As a consequence, literature and art have been the primary venues through which the Indigenous peoples of the Americas could express their traditional values and philosophy. As I have argued in previous chapters, much of the philosophy of these peoples is concerned with how to survive and prosper while living on intimate terms with the natural world and its nonhuman inhabitants, it seems logical to explore how this philosophy is exemplified and explained through the works of Indigenous American writers.

One category of literature is the "philosophical novel," defined as "that subspecies of fiction which endeavors to present a specific philosophical viewpoint, sometimes metaphysical, sometimes ethical, and sometimes aesthetic . . . whereas it is usually a defect in a work of fiction to ally itself closely with a particular viewpoint, for a philosophical novel, a grasp of the fact that a particular world-view is embodied is a precondition of understanding the novel" (Honderich 2005). Following this definition, I suspect that most important works of fiction by Indigenous writers fall within this category, even though they are not arguing philosophy in the Western sense. Indigenous authors are attempting to reveal the Indigenous way of thought and life to readers who are typically unfamiliar with these perspectives and are "reinventing the enemy's language" (Harjo and Bird 1997) in an effort to keep a sense of Indigenous identity alive.

An example of this approach can be seen in this statement by the Mohawk author Beth Brant:

> We have been forced to reject and forget what makes us real as Native peoples. The dominant society longs for this forgetfulness on our

> parts, hungering for our assimilation into their world, their beliefs, their code. It hurries this process along by promises of acceptance and forgiveness. Their paranoia becomes our own. But what we may have "forgotten" remains in our blood. Salmon's desire to home is our desire also. Blue heron's desire to fly long distances to make a home is ours also. Corn's desire to grow is ours also. For we are parts of them and they are parts of us. This is why we are Indigenous . . . what none of us has truly forgotten. (Cited in Eigenbrod 2005: 78)

Another example, which makes its point through sly humor, can be seen in the writing of the Fort Hope Ojibwa Ruby Slipperjack: "I could hear a loon calling . . . somewhere out on the lake. . . . Another loon answered and soon they were carrying on like two old friends catching up on news. I looked at Ol' Jim and asked seriously, 'Why do people say the 'lonely call of the loon' or 'the eerie hoot of an owl' when I have never heard these things by themselves. There are always two or a whole bunch of them.' Ol' Jim chuckled and said, 'Well, when you don't understand the language, all the voices sound the same, don't they?" (Slipperjack 1992: 159).

Clever as this is, it echoes the idea that observers who interact over a long time with animals pick up subtle cues to behavior (Barsh 2000). The reference to "understanding the language" is both an ironic comment on Euro-American failure to pay attention as well as a reminder that Indigenous people see animals as people. This also reflects how Indigenous people augment their ability to monitor ecological processes by observing other species with more acute senses (Barsh 2000).

These quotes from Indigenous (First Nations) Canadian writers illustrate how the novel can be used to communicate how Indigenous people interact differently with nature than does what Brant refers to as "the dominant society." Similar approaches can also be seen in novels written by members of U.S. tribes.

INDIGENOUS LITERATURE

Although there are a number of earlier works by Indigenous writers (see Owens 1992a for a review), Indigenous American writing began to be taken seriously as literature when Kiowa writer Scott Momaday's *House Made of Dawn* won the Pulitzer Prize in 1969. Indigenous literature continued to develop with the publication of other important and award-winning works, including James Welch's (Blackfeet/Gros Ventre) *Winter in the Blood* in 1974, Leslie Marmon Silko's (Laguna Pueblo) *Ceremony* in 1977, and Louise Erdrich's (Turtle Mountain Chippewa) *Love Medicine* in 1984. The pride and power generated by these major works stimulated other Indigenous voices and allowed Native American literature to flower and come into its own in the late 1980s and 1990s.

Most authors who have been significant contributors to this literature are of mixed blood. It has been argued that the "mixed-blood literary movement is personal, invented, appropriated, and irrelevant to First Nation status in the United States," because this literature "is not generated from the inside of tribal culture," which supposedly renders it a "literary movement of disengagement" (Cook-Lynn 1996). Cook-Lynn is of the opinion that Native American literature should avoid assimilation, despair, escapism, or fantasy and become engaged in a "rebirth of native nations" and "development of worthy ideas, prophecies for a future in which we continue as tribal people who maintain the legacies of the past and a sense of optimism."

In marked contrast to Cook-Lynn, the Blackfeet/Gros Ventre author James Welch stated, "The people are going to be getting further and further away from their culture, so actually the reservation will be just a place to live. There will always be Indians, but they won't be very traditional . . . on these small reservations" (Alberts 1988). An interesting implication of Welch's statement is that traditions may be more easily maintained outside of reservation communities because mixed-bloods who live in rural areas manage to establish lives that are place based and Indigenous in philosophical outlook, but with greater control of their own lives than Indigenous people would have in either urban or reservation settings.

"If a writer happens to be of mixed ancestry, . . . aware of and interested in one or both sides of that mixture, and if the writer seeks understanding for language of himself and his reader by the telling of stories, should the writer not write of that 'mixed-blood' subject that is the complex self? . . . To do so, of course, is to write of those tensions, dynamics and torsions of family, history, community, and nation that give rise to the hybrid self" (Owens 1998: 153). Such questions lie at the center of all effective writing that deals with personal experience, and how such experience contributes to worldview.

One factor that renders this writing Indigenous is that in novels and stories written by these authors, humans almost always have good relationships with the natural world; unlike in novels written by people of European ancestry, nature is rarely if ever threatening and frightening. Humans are never endangered by their experiences with animals—the main threats to Indigenous protagonists come from either the dominant culture or from within their own Indigenous communities. In European literature nature is predominantly regarded as an enemy that needs to be conquered. If characters go out into nature, it is most always a place where bad and frightening things happen to humans. In Indigenous literature, connection to nature is seen as a way of restoring Indigenous identity and healing individuals who have been damaged by their interaction with the dominant culture.

My intention is to discuss a series of novels and thematically related essays by major Indigenous authors and see how they employ the natural world as an element in their writing. I do not write overall reviews of these novels. Many literary critics and academics have already done this. Instead

I simply discuss how these authors employ the nonhuman world in their stories and how this differs from other literary traditions. For any reader who wishes to read more detailed commentary and criticism from a literary perspective I recommend Owens (1992a) and the voluminous references cited therein as a good starting point.

I begin with Momaday because of his significance as the first widely read and recognized Indigenous novelist. I follow discussion of Momaday by discussing the work of Silko and Erdrich and close with an extensive discussion of the work of Owens himself. My choice of these four individual writers is not meant to imply that I feel these are the most significant writers, or that they exemplify my theme better than do others. In fact, Welch's novel *Fools Crow* (1986), one of the few novels by an Indigenous writer set during the nineteenth century, is filled with discussion of nonhumans and their roles in the spiritual and ceremonial lives of the Blackfeet people. Similarly, Gerald Vizenor's short novel *Dead Voices: Natural Agonies in the New World* (1992) tells a story of urban survivance (survival combined with resistance) from the point of view of an Indian woman who transforms herself spiritually into a bear in order to survive in an urban environment. This character, Bagese, then tells a series of stories from an Indigenous perspective that reveal how a variety of nonhuman creatures perceive the urban world.

I chose these writers because they represent four disparate Indigenous cultural traditions and biogeographical regions: the Great Plains (Momaday), the southwestern desert (Silko), the north woods (Erdrich), and the southeast and the "civilized tribes" (Owens). In addition, choosing these four writers allows me to review works by two Indigenous men and two Indigenous women and to discuss seminal Indigenous novels such as Momaday's *House Made of Dawn* (1968) and Silko's *Ceremony* (1977), while also reviewing the spiritual and intellectual descendants of these major works, who stretch the boundaries and expand the themes. Finally, these writers all began their careers as novelists in different decades: Momaday published his first novel, *House Made of Dawn*, in 1966, Silko published her first novel, *Ceremony*, in 1977, Erdrich published *Love Medicine* in 1984, and Owens published *Wolfsong* in 1991. Thus these four authors also represent a sort of thematic progression in the Indigenous novel.

N. Scott Momaday: The Role Model

In his Pulitzer Prize winning novel *House Made of Dawn* (henceforth *HMD*), Momaday's basic story line is that Abel, the protagonist, is an orphaned World War II veteran who has been scarred by his wartime experiences. After becoming a killer (of both human and nonhuman) and then almost being killed himself, Abel is healed by re-establishing ties, first to the nonhuman world and subsequently with the landscape of his people. The title of the novel alludes to a section of the Navajo Night Chant that uses the metaphor of a house to represent the idea that the true dwelling place of Indigenous people is the natural world. Early in the book there is an extensive passage in which Abel watches two Golden Eagles (*Aquila chrysaetos*) involved in a courtship ritual that would seem to show a link between Abel and the natural world. Abel is too damaged by violence, however, and he is unable to appreciate the meaning of this nonhuman ceremony. Making things even worse, Abel ends up killing one of the eagles in a disrespectful manner. This act demonstrates how disconnected Abel has become from his indigenous culture and presages many of the problems that Abel faces. He has allowed himself to become almost lost to his culture to the point where he kills another human whom he regards as evil because he does not understand that evil cannot be killed.

In the early sections of *HMD*, Abel's failure to understand the implications of human relationships with the natural world in his tribal traditions separates him from both his grandfather and his culture. *HMD* contains an odd obsession with "evil" that seems to be represented by the color white, and oddly enough, by serpents. This last image appears to be an anachronistic biblical reference out of a novel by someone of European ancestry. In the American Southwest, snakes are typically considered by Indigenous peoples to be sacred guardians of springs (Basso 1996), rather than representations of evil.

An important rite of passage in *HMD* occurs seven years later when Abel lies, severely beaten, next to the ocean in southern California. This section of the book employs a metaphor drawn from nature—the spawning run of small fishes (grunion, *Leuresthes tenuis*), triggered by the moon. In an odd fashion, grunion are used as a metaphor for helplessness and victimization. Actual grunion are anything but victims. They are beings caught up in spawning so thoroughly that they seem to ignore possible predators; however, the timing and location of their spawning has evolved to minimize predation. Grunion spawn on the beach so that they and their eggs are less vulnerable to more serious predators on eggs that live in the water. This is also why they spawn at night on only a few nights every year, which makes it hard for terrestrial predators other than humans to take advantage of their brief period of vulnerability because they swamp their predators through sheer numbers. In my research in Newfoundland I observed similar behavior in capelin, *Mallotus villosus*, a species that anchors the food chain in the North Atlantic (Pierotti 1988a; Pierotti and Annett 1990, 1991). In reality grunion and capelin are no more vulnerable than salmon making their spawning runs. They are not helpless, but they are an important food source to a number of other species, including humans. It appears that Momaday was more interested in grunion as a metaphor for helpless victimization than as actual living organisms.

Abel conflates his own helplessness and weakness with that of the fish, but he also realizes that the moon influences the fish and the sea and has a major impact on his own life. This evokes a reawakening of a sense of connectedness in the natural world, leading him to undergo a Navajo Night Chant, which employs the phrase that serves as the title of the book (Schubnell 1985). In his injured state he recalls a time when he collected a water bird he had shot, establishing a spiritual link between the bird and his own desperate state. After this experience, Abel returns to his tribe and undergoes the rituals that allow him to return to his people in a spiritual as well as physical sense.

HMD seems to be an odd hybrid of Western and Indigenous styles and themes. The authorial voice in *HMD* is found not in Abel, but in the Kiowa Tosamah, a Los Angeles preacher who speaks a garbled Christianity mixed with Indigenous subthemes (Owens 1992a). Tosamah is allegedly a trickster figure, but as a Christian preacher he also represents the colonizing power. Tosamah also tells his "flock" a version of stories that make up the opening section of Momaday's later work *The Way to Rainy Mountain* (1969).

There are themes that relate to nature in *HMD*, but they are mixed up and conflated with Christian imagery and other white fantasies. They also reflect an odd sort of hostility and apparent fearfulness toward nature: "Loneliness is there as an aspect of the land" (127). "The skyline in all directions is close at hand. . . . There is a perfect freedom in the mountains, but it *belongs to the eagle and the elk, the badger and the bear.* The Kiowas reckoned their stature by the distance they could see, and *they were bent and blind in the wilderness*" (130; emphasis added). These statements are odd and seem to be an attempt to invoke nature without really showing any understanding of what nature truly is or how Indigenous people have traditionally related to it. To begin with, badgers are not creatures of the mountains at all. Along with their friends the coyotes, badgers are the quintessential carnivore of the Great Plains. Momaday seems to associate bears and badgers with one another, because he also invokes this pairing in a white woman's fantasy about Abel, while she watches him chop wood early in *HMD*: "Once she had seen an animal slap at the water, a badger or a bear. . . . She would have liked to cup her hand to the wet black snout, to hold for a moment the hot blowing of the bear's life . . . He was there rearing above the wood" (32). In addition, there is the troubling use of the term *wilderness* in which the Kiowa were "bent and blind."

Metaphors and symbols in *HMD* seem drawn more from Christianity than from Indigenous belief systems: the supposedly helpless, victimized fish that are a clear reference to Christ, the use of serpents as a symbol of evil, and Momaday's decision to give characters names with obvious references to Christian rather than Indigenous stories. For example, Abel was the Bible's first victim, the son of Adam and Eve, murdered by his own brother (Schubnell 1985).

In Momaday's later works, however, links to nature and Indigenous concepts are more apparent. The most obviously nature oriented of his works is the story cycle and memoir *The Way to Rainy Mountain* (1969), in which Momaday mixes family history, tribal stories, and the landscape of the

Great Plains where the recent cultural and spiritual development of the Kiowa people (Momaday's principal tribe) took place. In *Rainy Mountain* Momaday seeks his spatial orientation by opening with a description of Rainy Mountain itself and closing with the oft-quoted statement, "Once in his life man ought to concentrate his mind upon the remembered earth . . . to give himself up to a particular landscape. . . . He ought to imagine the creatures there and all the faintest notions of the wind. He ought to recollect the glare of noon and all the colors of dawn and dusk" (Momaday 1969: 83).

There is a series of odd statements in the prologue to *Rainy Mountain.* Speaking of the Kiowa origin story, Momaday writes: "It was there, they say, that they entered the world through a hollow log. The end, too, was a struggle, and it was lost. The *young Plains culture of the Kiowas withered and died* like grass that is burned in the prairie wind. There came a day like destiny; in every direction, as far as the eye could see, carrion lay out in the land. The buffalo was the animal representation of the sun, the essential and sacrificial victim of the Sun Dance. *When the wild herds were destroyed, so too was the will of the Kiowa people, there was nothing to sustain them in spirit*" (Momaday 1969: 3; emphasis added).

This section is followed immediately by the statements cited above how the Kiowa "*were bent and blind in the wilderness.*" This account seems to reflect an Indigenous version of the Euro-American fantasy of all Indians as riding horseback and chasing buffalo. More seriously, it treats the culture of the Kiowa as being "lost" and having "withered and died." Such sentiments are the antithesis of contemporary Indigenous activism, which works to restore and retain cultural tradition.

Admittedly, this was written in the late 1960s, well before the Indigenous Renaissance of the late 1980s and 1990s. In the 1960s the termination period was just ending, and Indigenous activism was just developing (e.g., Deloria 1969). Still, the victimization and defeatist perspective contained in these statements seem to reflect the victimized perspective also seen in *HMD.* To put things in a proper perspective, around the same time Momaday was writing those words, Vine Deloria Jr. was recommending *Little Big Man* to people who wanted to understand "Indian attitudes towards life" (Deloria 1969: 16).

One important story told by Momaday in *Rainy Mountain* is the story of seven sisters and their brother, who changes into a bear. To save the sisters from their pursuing bear/brother the earth produced Rock-tree, the Devils Tower in eastern Wyoming, one of the most remarkable geological features in the world. Momaday's Kiowa name is linked to this geological feature and through it to the story of the boy and his sisters. Possibly as a consequence, in Momaday's second novel *The Ancient Child* (1989) the protagonist Locke Setman (Set) is linked to this story of the boy who turns into a bear. A West Coast artist and clear authorial stand-in, Set is drawn back into the Oklahoma plains of his people, where he is given the bear medicine bundle of the Kiowa. At the end of the book Set appears to reconnect to his tribal roots by literally reenacting this story and apparently changing into a bear.

Leslie Marmon Silko: Rediscovering Ancient Links

A decade after the publication of *House Made of Dawn*, Leslie Marmon Silko published a novel with a theme superficially similar to that found in Momaday's novel. The protagonist in *Ceremony* (1977) is a young, orphan mixed-blood from Laguna Pueblo, Tayo, who has been damaged both spiritually and psychologically by his experiences in the Second World War. *Ceremony* has the courage to challenge the myth of the veteran, the idea that Indigenous men (and women) who fight for the colonizer are the contemporary equivalent of warriors from the old times who resisted the invasion. In *Ceremony* the veterans (mostly pureblood) other than Tayo are involved in witchcraft, which explicitly comes from their association with white culture.

Many literary reviewers have written papers to analyze the complex themes and interweaving of stories in *Ceremony* (e.g., Owens 1992a and references therein). I concentrate my analysis solely on the use of nature in Silko's writing to indicate how the land and nonhumans persist and survive despite the devastation brought on by the actions of Western culture and politics.

The importance of the natural world is evident at the very beginning of *Ceremony*, with Thought-Woman, the spider, spinning the web of the story. Almost immediately we learn that Tayo had "to sweat to think of something . . . that existed by itself, standing alone like a deer. And if he could hold that image of the deer in his mind long enough his stomach might shiver less and let him sleep for a while" (Silko 1977: 7). Shortly thereafter, Silko weaves in a story where Corn Woman scolds her sister Reed Woman for bathing too much, and Reed Woman departs, taking away the rain with her (13). This story parallels Tayo's experiences with traditional stories of the Keresan peoples.

As Tayo travels to a bar with another veteran who drinks excessively, he thinks about animals "and the way they drifted with the wind . . . only humans had to endure anything because only humans resisted what they saw outside themselves. Animals did not resist . . . they persisted, because they became part of the wind" (27).

The narrative in *Ceremony* is nonlinear—A series of stories and themes weave around the experiences of Tayo, who is convinced that his prayers in the Philippine jungle for the rain to stop during the war resulted in a drought in his homeland. A healer called to work with Tayo explains to him that "'you know grandson, this world is fragile.' . . . The word he chose to express 'fragile' was filled with *the intricacies of a continuing process*, and with a strength inherent in spider webs woven across paths. . . . It took a long time to explain the fragility and intricacy because no word exists alone. . . . That was the responsibility that went with being human" (35, emphasis added). Later on Tayo recalls a visit to a spring with his uncle Josiah: "nobody could ever remember a time when the spring had dried up. . . . 'You see,' Josiah had said . . . 'there are some things worth more than money. . . . This is where we come from, see. This sand, this stone, these trees, the vines, the wildflowers. This earth keeps us going. . . . These dry years you hear . . . the wind and the dust, they are part of life, too like the sun and the sky. You don't swear at them. It's people too, see. They're the ones' . . . Tayo . . . knelt . . . closed his eyes . . . and tasted the deep heartrock of the earth, where the water came from, and he thought maybe this wasn't the end after all" (45–46). This alludes to a parallel story of how Hummingbird and Big Green Fly must travel through the layers of the world to get Corn Woman to allow the water to return to the world.

There are numerous themes and stories in *Ceremony*, but it is clear that the physical environment and the nonhumans are constant and important actors in these stories. Tayo is slowly restored to spiritual health through continued re-exposure to the landscape of his people and the cultural meanings that emerge from this place where his people came to be. The core of Tayo's trauma lies in the death of his cousin/brother Rocky, a full-blood, who we are repeatedly told "deliberately avoided the old time ways," and who took the life of a deer without showing proper respect. "They sprinkled cornmeal on the deer's nose and fed the deer's spirit. They had to show their love for the deer to show their love and respect; their appreciation; otherwise the deer would be offended, and they would not come to die for them the following year" (51), however, Rocky "was embarrassed at what they did." Despite living on a reservation, Rocky became highly assimilated, apparently with the encouragement of his full-blood mother. One hidden subtheme in *Ceremony* is that what ultimately may save Tayo from witchcraft is that he was not fully accepted by his assimilated full-blood aunt and cousin. The full-blood veterans seem to have turned into witches, and his cousin Rocky was so disconnected from the earth that he could not survive the war. The contaminated state of the full-bloods of Tayo's generation echoes the quotation from Welch (above) about reservation communities.

When Josiah decided to go into the cattle business, the eighteen-year-old Rocky tells them about books provided by the local agricultural agent. "*Those books are written by scientists . . . (who) know everything there is to know* about beef cattle. That's the trouble with the way the people around here [Keresan Pueblo] have always done things—they never knew what they were doing" (76, emphasis added). Over time it becomes clear that Rocky's rejection of his Keresan identity, especially his lack of respect for the nonhuman world, ultimately led to his death, a theme we also see developed in other novels by Indigenous authors.

The last third of *Ceremony* represents powerful storytelling, and it is a story that could only be told by an Indigenous author. Tayo recovers enough to go looking for his uncle's cattle that were stolen while he was away at war. First he meets a woman, T'seh Montano. Owens (1992a) suggests that she is the rain spirit of the sacred mountain Tes-pi'na (Mt. Taylor to the whites). However, something T'seh tells Tayo suggests something even more interesting than the Owens' interpretation. Tayo says, "You never told me your name." She responds, "I'm a Montano, you can call me T'seh. That's my nickname because my Indian name is too long" (223). To me, this suggests that she is the spirit of Tes'pi'na itself; after all, she is a Montano.

T'seh's love for Tayo and his love in turn for her restore him further. Tayo has gone to the slopes of the mountain to heal himself. However, initially when Tayo goes after the cattle he feels despair. At his lowest moment Mountain Lion (the hunter) appears to remind him of his responsibilities to the earth. "Tayo got to his knees slowly and held out his hand. 'Mountain lion' he whispered, 'becoming what you are with each breath, your substance changing with the earth and the sky.' The lion blinked his eyes; there was no fear. He gazed at him for another instant . . . and then disappeared into the trees" (95–96). In contrast with Western writing, in which an encounter with a cougar would be frightening (e.g., Baron 2004), this experience reduces Tayo's fear and feeling of helplessness, and enables him to go on. This is reinforced after he is thrown from his horse and captured by white range-riders. Mountain Lion reappears and leads away the whites who prefer to hunt the hunter. As Tayo walks through the snow he encounters Mountain Lion in his human form as the hunter, singing a chant to call the deer. The hunter feeds Tayo and takes him back to the cabin of T'seh, where he finds his cattle, who are being held for him by these spirits of the mountain.

Much more follows, but from the perspective of this chapter the important theme is that Tayo's respect for and reconnection to the nonhumans and the physical environment help him end the drought and reconnect with his people through his links and caring for the place where they came into being.

As Momaday did in *The Way to Rainy Mountain*, Silko expanded on the themes found in *Ceremony* in her combined memoir, collection of essays, and stories from her people, *Yellow Woman and the Beauty of the Spirit* (1996). The heart of this book is in the following quote, which incorporates themes from other chapters in my book:

> The antelope merely consents to return home with the hunter. *All phases of the hunt are conducted with love*: the love the hunter and the people have for the Antelope People, and the love of the antelope who agree to give up their meat and blood so that human beings will not starve. *Waste of meat or even the thoughtless handling of bones*

cooked bare will offend the antelope spirits. Next year the hunters will vainly search the dry plains for antelope. Thus it is necessary to return carefully the bones and hair and the stalks and leaves to the earth, who first created them. The spirits remain close by. They do not leave us.

The dead become dust, and in this becoming they are once more joined with the Mother. The ancient Pueblo people *called the earth the Mother Creator of all things in this world.* Her sister, the Corn Mother, occasionally merges with her because all succulent green life rises out of the earth. (Silko 1977: 26–27; emphasis added)

Silko appears much more confident than Momaday in the way she writes about the natural world, the relationships between humans and nonhumans, and the spirits that help link the two. Momaday appeared to grow more confident over time, as *Ancient Child* reveals much greater comfort with nature and is less filled with fear and self-loathing than is *HMD*. Still, the way Silko writes of her places and her people's relationship to those places gives much more of a feel for the land and its meaning.

Louise Erdrich: The Nurturing Spirit of the Land

This feeling of comfort with a place and its meanings are evident in the writings of Louise Erdrich, who is probably the most commercially successful Indigenous writer. Erdrich writes primarily about the multigenerational family she has created, which includes both Chippewas and Euro-Americans of German ancestry, reflecting her own ancestry. Erdrich is a very "female" writer—few write more eloquently or expressively about the costs and benefits of female passion, both physical and spiritual. Her spiritual power allows her to invest her stories with the feel of a world that exists alongside the physical one but is not truly separate from the physical world.

Erdrich draws naturally upon images and themes from nature. Her third novel, *Tracks* (1988), takes as its theme shape-shifting and the importance of traditional stories and values among the Chippewa during the period following the Dawes Act and Indian allotment, when whites were trying to take over rich timber resources on her reservation. This novel introduces Nanapush, who is both an elder and Erdrich's interpretation of the Anishinaabeg trickster figure Nanabozho (Owens 1992a). Another major character is Fleur Pillager, the only survivor of an epidemic that wiped out her family and a woman who lives close to the nonhuman and spirit worlds. The title refers to both tracks left by living beings as they move across the land and the letters written on paper that are used to steal land and natural resources from the Chippewa.

Erdrich's nonfiction memoir of one of her pregnancies, *The Blue Jay's Dance* (1995) reveals the desire of life to exist despite the dangers of the world. This book takes its title from an observation Erdrich made of a Blue Jay (*Cyanocitta cristata*) standing off an attack by a Red-tailed Hawk

(*Buteo jamacensis*) by leaping and twisting itself in front of the startled predator, who abandoned its pursuit in the face of this defiant dance. I lived with a Blue Jay for eleven years, and I know exactly how defiant these small but powerful entities can be in forcing their will upon a situation. I also know that they truly do dance.

Erdrich achieves flights of lyrical imagery in her storytelling, especially when describing the natural world. Consider her description of the experience of a thrice-orphaned girl from *The Antelope Wife* (1998):

> The antelope emerge from the band of light at the world's edge. A small herd of sixteen or twenty flickers into view. Fascinated they poise to watch the girl's hand in its white sleeve dip. Feed herself. Dip. They step closer. Hooves of polished metal. Ears like tuning forks. Black prongs and velvet. They watch Matilda. Blue Prairie Woman's Daughter. Other Side of the Earth. Nameless.
>
> She is seven years old. . . . She doesn't know what they are, the beings, dreamlike, summoned by her [dying] mother's song, her dipping hand. They come closer, closer, grazing near. Folding their legs under them to warily rest. . . . In the morning, when she wakens, still holding her mother's hand, they are standing all around. They bend to her, huff in excitement when she rises and stands among them quiet and wondering. Easy with their dainty precision, she wanders along in their company. Always on the move. . . . The herd flows in steps and spurting gallops deeper into the west. When they walk, she walks, following. . . . When they run, she runs with them." (19–20)

This passage captures the essence of pronghorn antelope (*Antilocapra americana*) and their unique combination of curiosity, speed, and power. This species is very fast, but also very curious and trusting. At the same time we get the feeling of being in the presence of something that seems outside of normal human experience, certainly of Euro-American experience.

Even more powerful is Erdrich's description of a buffalo (or bison) hunt in *The Last Report on the Miracles at Little No Horse* (2001), Erdrich's complex novel that deals with the impact of Europeans on the Chippewa people and vice versa. This is probably one of the most powerful images ever generated on the destruction of the buffalo herds by Euro-Americans.

> (In the wake of this hunt) the herd did not bolt away but behaved in a chilling fashion . . . the surviving buffalo milled at the outside of the carnage, not grazing, but watching with an insane intensity. . . . Even through the night, the buffalo stayed, and were seen by the uneasy hunters and their families to have remained standing quietly as though mourning their young and their dead, all their relatives that lay before them. . . . At the sun's zenith . . . the buffalo began to make a sound. It was a sound never heard before, no buffalo had ever made

> this sound. . . . An unmistakable and violent grief, it was as though the earth itself was sobbing. . . ."The buffalo were taking leave of the earth and all they loved" said the old chiefs and hunters. . . . "The buffalo went crazy with grief to see the end of things. Like us, they saw the end of things, and like many of us today, they did not care to live." (Erdrich 2001: 157–158)

This story illustrates the multidisciplinary aspect of Indigenous culture and thought and how these can be revealed through fiction. In televised interviews Erdrich has said that this account is based on actual historical records, so this story incorporates literature, history, and animal social behavior. It is crucial to remember, however, that this account describes the behavior of a group of nonhumans, but the behavior described is one that the Western scientific tradition is incapable of explaining, also a behavior they probably would not acknowledge even if they could explain it, because it is a unique event and will never be replicated. Given the emphasis placed by Indigenous cultures on relationships and behavior (Barsh 2000), for the Anishinaabeg to witness this must have been nearly unendurable, because this reflected their own experiences with both Europeans and the diseases they brought with them. I cannot read this account without a major emotional response. (The version I have provided is edited—to experience the full impact the reader should see the original.)

Louis Owens: The Importance of Place

The final author I discuss is the late Louis Owens (Choctaw/Cherokee/Irish). Owens incorporated the physical landscape and the nonhuman elements of the community as vital presences more than any other writer. Place is very important in the identity of Indigenous people, and no writer, of any ancestry, is better than Owens at emphasizing how place figures into the identity of his Indigenous characters. "I think *my works are about the natural world and our relations with that world*. . . . Though *each of my novels begins and ends with place itself*, the mysteries of mixed identity and conflicted stories, both the stories we tell ourselves and the stories others tell about us and to us, are what haunt my fiction" (Owens 1998: 181; emphasis added).

One reason Owens is effective in writing about place is that he sets his novels in localities that were important and very familiar to him. Owens reveals these locations as sacred landscapes. The Stillaguamish River valley and the North Cascades are the setting for *Wolfsong* (1991), Owens's first novel. It is also the place where Owens worked after completing his undergraduate degree (Owens 2001). Parts of the opening pages of *Wolfsong* are almost identical, word for word, with Owens's description of his first arrival in northern Washington (cf. pp. 59–60 of Owens's [2001] essay "In the Service of Forests" with pp. 13–16 of *Wolfsong*). Such a device was also used by Momaday, who incorporated much of the opening section of *Rainy Mountain* into *House Made of Dawn* (see above).

The southern Salinas Valley and California Coast Range around Atascadero, California, where Owens spent his teenage years, along with the Yazoo River country in Mississippi, where Owens was a young boy, are the setting for his second novel, *The Sharpest Sight* (1992b). The university campus and the coastal mountains around Santa Cruz, California, where Owens lived while he was a faculty member from 1990 to 1994 are the setting for *Bone Game* (1994). The mountains and desert around Socorro and Magdalena, New Mexico, where Owens hunted and fished when he was a faculty member at the University of New Mexico, play a similar role in *Nightland* (1996). I have lived and spent considerable time in New Mexico, the Salinas Valley, and the Santa Cruz Mountains (I was a young boy in New Mexico, and like Owens I was a teenager in coastal California). Owens evokes all of these landscapes fully as sacred spaces, making their physical characteristics and how people live in them an important emotional and spiritual component of his storytelling.

As a spiritual heir of Momaday and Silko, Owens employed the impact of war and the ghost sickness induced by war in Indian veterans as an important theme. Being their heir, he employed the war of his time, Vietnam, rather than Italy and the Philippines during the Second World War. Like Abel in *House* and Tayo in *Ceremony*, Attis McCurtain in *The Sharpest Sight* and Jacob Nashoba in *Dark River* are Vietnam veterans irrevocably damaged by their experiences and the racist attitude they perceived to be endemic in the U.S. war effort, which is also an important aspect of Silko's *Ceremony*.

As for nonhumans, *Nashoba* means wolf in the Choctaw language, and wolves are prominent presences in Owens's first novel, *Wolfsong*, and his last novel, *Dark River* (1999). The wolf familiar to the Choctaw and Cherokee is not the group-hunting, pack-forming gray wolf familiar to most tribes, but the smaller, solitary-hunting red wolf (*Canis rufus*) of the bottomland hardwood forests of the South. As a Choctaw, Jacob represents wolf as both a warrior and an excellent hunter, but like the red wolf, he also moves and acts alone. In contrast, the symbolic role of the gray wolf in Plains culture is as a good hunter and provider who is connected to and part of a large family, and who taught Indians to hunt and hunted with them (Schlesier, 1987; J. Marshall 1995).

Wolf serves as an important link to the natural world for Owens. In *Wolfsong* (1991) Owens's protagonist, Tom Joseph, is the spiritual heir to his uncle Jim, who received Wolf as his guardian spirit during his vision quest. Jim Joseph actively resisted the development of a wilderness area and river valley that he regarded as sacred, and is defending this place when his spirit is called home by wolves and by "drummers and dancers." Tom is attended regularly by a wolf, presumably his uncle in spirit form, which reveals the significance of the title. Rather than reflecting a tribal activity, as in the titles of *House Made of Dawn* and *Ceremony*, Owens's title reflects the rituals of the nonhuman world.

In exploring the theme of what it meant to be mixed-blood in late twentieth-century America, Owens created protagonists who are generally relatively passive, conflicted, and of mixed-blood ancestry. The elders in Owens's novels are compelling and powerful as characters. They advise, try to protect the younger characters, and serve a role similar to that of Thought Woman in *Ceremony*. The difference between Owens and Silko is that Owens's elders are not metaphorical, but literal, because Owens treated nature and biological phenomena as direct actors in his stories. In wolf form, Jim Joseph draws his nephew Tom more deeply into the life of the mountains, rivers, glaciers, and forests of the North Cascades.

Unlike with Abel in *House*, or even Tayo in *Ceremony*, none of Owens's Indian characters are victims, even when they suffer discrimination and race-based harassment. They are only lost if they lose touch with their tribal identities, as was also the case with Rocky and other veterans in *Ceremony*. Owens does not automatically treat Indians as saints and whites as sinners. In *Wolfsong*, white characters are troubled by environmental destruction, yet Owens realized that the major difference is that whites accept "progress" and development as inevitable outcomes of their way of life. The fear with which the dominant culture regards what Gerald Vizenor calls "survivance" on the part of Indians generates many of the conflicts.

Tom's decision to continue his uncle's survivance leads him to resist development of the valley and a copper mine, he is pursued through the mountains by whites, while he is attended and guarded by wolves as he makes his escape. Another key element is that during all of their attempts to stop development, both Jim and Tom Joseph are accompanied by ravens who observe and comment on the action. This has been interpreted as an effort by Owens to introduce "trickster" figures into his narrative (LaLonde 2002). Although Raven undoubtedly functions as both trickster and creator in Salishan stories, Owens also describes a well-established major ecological relationship in ravens' accompaniment of wolves in their activities, serving both as an early warning system and locators of food. These two species have a close and mutualistic relationship in North America (Heinrich 2000). As an accomplished naturalist Owens was surely aware of this relationship and built it into his story. Ravens accompany and warn Jim and Tom as the human heirs of the wolf spirit.

Owens followed *Wolfsong* with *The Sharpest Sight* (1992b), in which he is fully in command of the spiritual aspects of both nature and traditional stories. Only Erdrich is as successful as Owens at fitting characters from tribal stories and traditions into contemporary narratives. *The Sharpest Sight* introduces Cole McCurtain, a mixed-blood Choctaw and Owens's most autobiographical character. Underlying this surface story is the theme of Choctaw culture and the responsibility that Choctaws owe to family members of all kinds. Cole flees California to avoid the draft and hides out with his "uncle" Luther Cole in the Yazoo River country of Mississippi. The FBI follows him there in an effort to arrest him for draft evasion. Overseeing these events are a number of elders, both living and nonliving, who attempt to protect and guide their descendants, and who also provide an overarching historical perspective. Owens tells the story in a nonlinear fashion, which loops back through the history of events and their subsequent consequences, involving local Indians, whites, and Hispanics.

In Mississippi, Cole learns a great deal about what it means to be Chata (Choctaw) through his interactions with his uncle Luther Cole and Onatima Blue Wood, both mixed-blood Chata elders. These sections show a strong Indigenous combination of humor and power and might be criticized as escapism or "fantasy" (Cook-Lynn 1996), however, they go to the essence of tribal identity and reveal that tribal elders are powerful in ways completely separate from the political and economic power wielded by whites. Learning how to live properly from these elders heals and strengthens the confused Cole, while at the same time providing him with a sense of tribal identity.

In *Wolfsong* the animal spirit featured prominently was wolf, who functioned as a supportive guardian and teacher. In *Sharpest Sight* the animal power spirit is *koi* or *Nalusichito*, the black spirit panther that appears on the bridge over the Salinas to alert Hispanic deputy Mundo Morales about Attis's body in the river, and then reappears in the Mississippi swamps. Luther has brought Attis's *shilombish* (ghost) back to his cabin, where it waits for its remains. The matter-of-fact manner in which Luther and Onatima deal with these extraordinary events illustrates the power of traditional tribal life without being mystical or metaphorical. *Nalusichito*, the soul catcher (Owens 2001), was aroused by Attis's death. Both it and the *shilombish* can only be appeased by the return of Attis's bones to Mississippi, which must be done by Cole and their father Hoey. The *Nalusichito* sequence is based on an actual experience Owens and his family had in his childhood in Mississippi, when a panther followed his father home and jumped on the roof of their cabin (Owens 2001).

Like *Wolfsong, Sharpest Sight* is a complex, multifaceted story involving Indian people, who, although not powerful in the sense of the dominant culture, are perfectly capable of taking care of themselves in extreme circumstances, and who draw much of their integrity from their tribal identities and the traditions and beliefs of their peoples. The full-bloods Jim and Tom Joseph in *Wolfsong* and the mixed-bloods Hoey, Luther, and Onatima in *Sharpest Sight* are smart, competent people who draw their strength from their understanding of the land and what they know of their cultural traditions. In contrast, the white characters are conflicted, and their attempts to control situations only succeed in provoking violence and destroying their own need for peace and economic (and psychological) security. The Indian and Hispanic elders in *Sharpest Sight* are wise and empowered, and their primary goal is to assure that their descendants remain as protected as possible from the destructive forces that have been set loose, while remaining aware of the traditions and connections that link them to the physical places from which they come. Along with Erdrich's characters, this represents obvious progress from the frightened, victimized Abel of *House Made of Dawn* in the way Indigenous characters can be portrayed.

In *Bone Game* (1994) Owens returned to the story of Cole McCurtain 25 years after the end of *Sharpest Sight*. Cole is now a respected Indigenous professor, however, unbeknownst to him, some of his white male students are obsessed with Indian "traditions" and with him as a "genuine Indian instructor." They have been killing other whites in some sort of unspecified ritual sacrifice, which links them in a synergistic interaction with the vengeful spirit of Venancio Asisara, an Ohlone spiritual leader. Asisara is "a consciousness awakening and looking around at a world entirely familiar and entirely alien simultaneously, rather the way the Ohlone were forced to see in a single generation the near destruction of their cultural world, by the Spanish" (L. Owens, personal communication). Monterey Bay is still the Ohlone homeland, but it has been made alien by urban sprawl and habitat destruction.

In his critical study of Owens's novels LaLonde (2002) characterizes Asisara as the embodiment of evil, and Venuto (1998: 39) argues that "Venancio's continued presence demonstrates that evil, which will always exist, must be acknowledged in order to maintain the world's balance." These are interpretations from the dominant culture. Owens described Asisara as follows: "My sense is one of immense anger and power, the kind of power that comes with being a holy person, but warped into an uncontrolled anger that rises out of the very earth itself, set in motion, or awakened, by the violence of the two murderers" (Louis Owens, personal communication).

As in *Wolfsong* and *Sharpest Sight*, in *Bone Game* a powerful animal spirit is present, the grizzly bear, who appears with, or as, Asisara. Like the wolf in the North Cascades, the grizzly bear is now extinct in California, yet their spirits are powerful symbols and very familiar to the Indigenous peoples of those locations. The combination of Asisara and the grizzly represents conjunction of two powerful spirits that have been exterminated from their homeland, yet whose angry spiritual power is being tapped by the two murderers.

Nightland (1996) is set in the mountains and desert around Magdalena, New Mexico, and Cibola National Forest, and deals with drug dealing and double-crossing driven by historical desires for vengeance and retribution among Southwestern peoples. Caught up in this chaos are two mixed-blood Cherokees, Billy Keene (Kaneequayokee) and Will Striker (Manstriker). These two cousins, given the same first name, are the children of Cherokee men who married white women and moved from Oklahoma to New Mexico to escape racial prejudice directed at mixed marriages. They share ranchland purchased from a New Mexican Hispanic family land grant, which contains a spring where the Spanish massacred a group of Apaches.

The two men have grown up virtually as brothers; however, like Tayo and Rocky in Silko's *Ceremony* (1977), they are very different in personality. Billy is impulsive, outgoing, and identifies with the dominant culture. He has sold all his cattle and keeps no other animals. In contrast, Will is introspective and solitary—he has kept a community of nonhumans together, caring for an old dog, pot-bellied pig, two aging bulls, and an old horse, and allows his heifers to run free on his land for the remainder of their lives. Billy lives with "Grandpa" Siquani, a Cherokee elder who not only remembers living in Oklahoma but also participated in the Trail of Tears and experienced the lives of the Cherokee before the arrival of whites.

Nightland refers to the "land to the west" where the sun goes in the evening and where we all must someday follow, the Cherokee land of the dead. Will and Billy are deer hunting when a man and a suitcase containing nearly a million dollars fall out of the Western sky. Billy claims the money is a gift from the Great Spirit, whereas Will calls it "corpse money" and wonders what Siquani would think about this and what to do about the man's body, impaled on the top of a juniper.

There are numerous vivid characters, including Duane Scales, a ruthless white hit-man who is also a thoughtful if angry environmentalist, and Arturo Cruz, the Pueblo Indian who fell from the sky. After his passing, Arturo joins Siquani to discuss the implications of the action that transpires. As Siquani tells it, Billy's father (his descendant) asked him to come along when they left Oklahoma and he agreed to "see how the story ends." Siquani places things in perspective for Will, but he cannot save Billy from his folly, his greed, and his desire to "be American."

When Billy and Will return to their ranches, Will finds a number of buzzards (Turkey Vultures, *Cathartes aura*) hanging about his ranch. Billy finds a large number of crows at his place. Siquani explains that "Old Buzzard was real important back in the beginning. . . . Buzzard flew around to see where animals could live" (48). Buzzard is also the purifier to whom Hummingbird and Big Fly must offer tobacco in *Ceremony*. The buzzards see the difference between Will's place and Billy's. Will's place is "where the animals can live." Billy has given up on his place, his animals, and his people. Will's ranch represents the earth itself, cared for so the animals can live there.

In *Nightland* Owens expects Americans to face the consequences of their actions and stop denying that Wounded Knee and the Trail of Tears were acts of genocide, but he also shows how destructive Indigenous acts of reprisal can be. Siquani conducts rituals and ceremonies to protect Billy, who he knows is in great danger. Billy is too determined to "be an American." To survive, however, he needs to acknowledge his Indian heritage. Billy is seduced and killed by the full-blood Apache Odessa Whitehawk, who echoes the destructive full-blood witches in *Ceremony*. Will, however, is saved from Odessa by his nonhuman community.

At the end, Siquani conducts a Cherokee water renewal ceremony, which brings the rains back and restores the spring. As Siquani puts it, "Them coyotes and crows been having a big celebration all night and done wore me out." Will's animals kept him from being alone when he lived without human company, and his responsibilities to them kept him functioning during these events. When Will and Siquani meet at the finish, Siquani explains, "I had to help Billy find the path. . . . They been piling up things to hide it for all these years now, but our world is still here, Willum. Sometimes we forget because we got to look so hard to see it and people get tired . . . *but the animals know and they don't forget. We got to listen*" (214; emphasis added). The animals had been trying to warn Will and Billy all along. Buzzards came to Will's ranch because that was where the animals could live, and the crows kept flying over Billy and his ranch and trying to warn him, but he had lost the path and his connections to the living world, which is how he became "American" (Pierotti 2005b).

Dark River (1999) is different from Owens's other books in that his protagonist Jacob Nashoba is alone with no brothers or close friends. Jake finds the communal Apache lifestyle hard to cope with, and his ghost sickness from his Vietnam experiences frightens his Apache ex-wife. Like Billy in *Nightland* and Rocky in *Ceremony*, Jake has rejected his Choctaw heritage. An Apache elder confronts Jake about his heritage by asking, "Why didn't you ever tell your wife or her children about your own culture? Did you forget everything, or don't those Choctaw people you come from have stories?" His response is "Look Grandmother, I'm No Indian. My people have stories of leprechauns and something called the *sidhe*. . . . Besides home is where the heart is." This reference to Jake's (and Owens's) Irish heritage is a deflection, acknowledging the spirits and stories of a foreign land, while refusing to acknowledge the spirits and stories of his native land. As with other Indigenous characters who reject their heritage, this refusal is fatal.

Dark River develops themes hinted at in Owens's earlier work. In *Nightland*, Odessa asks Billy during an early stage of their relationship, "Have you ever wondered what it would have been like if Europeans had never come?" Billy watches a coyote walk through the sagebrush and responds, "Maybe we'd be a theme park, a kind of ocean to ocean Disneyland, where we could charge Frenchmen and Germans $5,000 for a weekend vision quest, $30,000 for a shaman school diploma." In *Dark River* the Apaches make money off a casino and sell hunts for trophy elk they raise on the reservation. An anthropologist who has come to live with the Apache, and who lives more traditionally than do the Indians themselves, asks to speak to the tribal council, contending that "there is a cultural cost involved in such enterprises," and suggests that they abandon their casino and lodge and become a traditional tribe, "tear down the government shacks and build authentic wickiups, eat venison and dried chokecherries . . . all the things our ancestors did." If they live traditionally, they could get grants from an incredible number of sources. An Apache elder argues that "it's not just the grants but all the tourists that'll come. . . . Nobody's going to go out to Sioux country because they get depressed by the poverty, and that bunch of Cherokees in North Carolina have nothing but a little circus and they still attract millions of tourist dollars. . . . We'd be bigger than Yellowstone or Yosemite, bigger than Universal Studios or Disney World" (72). One council member responds, "What're we going to do with all that money? I mean we can't have no satellite dish and color TV and no Land Cruiser." Another says, "We already have that Native American village thing in June with *authentic dwellings and dancers and stuff for the tourists*" (73; emphasis added).

One Apache follows up on this suggestion. Jessie, a young man who tries to use Jake as a role model, runs weekend vision quests for Europeans and uses the proceeds to start a scholarship fund. One of Jessie's fake vision quests turns into a genuine spiritual experience, which only Jake and Jessie can resolve. Jessie has taken his client, a Frenchwoman, into the canyon for her vision quest. Wearing a realistic wolf costume to give his clientele a "genuine" vision, Jessie is shot and killed by militia who think they are rescuing his client from attack. The militia hold the client and Jake's granddaughter, Alison, who was also in the canyon on her own personal vision quest. Jake goes into the canyon to rescue Alison, and Jessie returns in the actual form of a wolf.

Dark River descends quickly into violence. The militia are all killed by Jake or by the women with the help of Jessie as a spirit wolf. In all of this violence Jake is mortally wounded and falls into a hole in the earth. Exactly how *Dark River* ends is unclear and this hints of the ambiguity present in Owens's own life near its end. Owens posits several endings and then has them rejected by various characters as overly clichéd, too postmodern, unacceptable to Mrs. Edwards, and so on. All that seems clear is that Jake is attended by Grandmother Spider, maker of the story in *Ceremony*, and also the helper of the Apache hero Monster Slayer. Grandmother Spider takes Jake home into the earth. After his passing Jessie changes from spirit helper into a proper ghost and leaves as well.

What is both fascinating and troubling is that Owens started *Wolfsong* with Jim Joseph being called home by wolves and traditional drumming as his spirit passed in the woods. Now with Owens's death, his last novel ends with Jacob Nashoba falling into the earth. Jake's last name means wolf. "To an Apache the coyote might have been disturbing. If they knew what Nashoba meant, they'd be even more disturbed by him, too" (124).

The arc of Louis Owens novels starts with a wolf as the symbol of Indian survivance and ends with the passing of a valiant Indian warrior who is also identified with the wolf, linking these characters in different novels at a spiritual level. No Indian writer has been as effective at illustrating and describing the true sense of Indigenous spirituality (Pierotti 2005b). Owens's passing has robbed all of Native America of one of our truest voices, even though it seems almost foreshadowed in his own writing because Jacob Nashoba is also the first of Owens's protagonists to die.

There is also an arc to the history of the Indigenous novel. Momaday may appear less sophisticated in his writing about Indigenous spirituality and connections to the natural world, but he was the first to seriously address these issues, blazing a trail that others followed. Like Momaday, James Welch did not deal effectively with spirituality and nature in his first novel, *Winter in the Blood* (1974). By the time he wrote his third novel, *Fools Crow* (1986), Welch had grown in confidence and thoughtfully addressed the issues of relationships with nature and spirituality.

Silko followed the trail blazed by Momaday in writing *Ceremony*, telling a story with superficially similar themes in a much more Indigenous fashion, employing themes that connected directly to the stories of her Keresan people, rather than relying on family history and a generally vague understanding of Kiowa tradition, as did Momaday. Silko wrote one of the strongest Indigenous novels ever produced, but she was able to follow the trail blazed by Momaday.

This first generation of contemporary Indigenous writers (it should be emphasized that Momaday and Silko are still active and productive) created a tradition that allowed the next generation of writers like Erdrich and Owens to grow in confidence and expand their themes, writing confidently from an Indigenous perspective. Erdrich has created a multigenerational complex Indigenous community that she visits from a variety of perspectives, telling stories of both Indigenous and Euro-American people, even including both in the same novel. Erdrich writes primarily about the human stories, yet as I have shown, she can write powerfully from the perspective of nonhumans, telling stories that no one else has dared tell. Louise Erdrich has taken the Indigenous novel and transformed it through her imagery and language.

Finally, Louis Owens built on the work of all of his predecessors to create a new type of Indigenous novel. Owens understood the height of the shoulders upon which he stood, having written an excellent history of Indigenous novels (Owens 1992a), in which he not only persuasively argued for the importance of Momaday, Welch, Silko, and Erdrich, but also critically reviewed the work of early Indigenous writers, such as Mourning Dove (Cogewea), John Rollin Ridge, and Darcy McNickle.

Owens took a genre, the thriller, and turned it into an Indigenous genre by standing it on its head. In *Sharpest Sight, Bone Game, Nightland*, and *Dark River* Owens created situations in which Indigenous people who maintained their traditional values and knowledge were the ones who survived to carry on, whereas numerous Euro-American as well as Indigenous characters who tried to assume a European/American identity suffered the consequences of their actions. Spirits, and predatory animals, typically seen as dangerous and frightening from the European perspective, are not dangerous in Owens's stories (Pierotti 2005b). In many cases they are helpful to their Indigenous relatives. In doing this Owens reversed the moral and ethical assumptions of the Western world and established that an

Indigenous way of viewing the world, combined with a knowledge of place, allows its adherents to survive the most dangerous aspects of our modern world, including psychotic killers, drug cartels, and armed right-wing militias. Following the tradition established by Momaday and modeled by Silko, Erdrich and Owens have succeeded in telling new stories that are truly empowering and decolonizing: Erdrich through emotional ties among people and Owens through spiritual links to place and nonhumans.

8 Ecological Indians

European Imaginations and Indigenous Reality[1]

"... we all learn to accept that most Americans speak double-speak, a language that—like all dystopian languages, including those we read about during adolescence in Aldous Huxley and George Orwell—obscures, veils, rewrites, self-justifies."

Patricia Penn Hilden (1995: 78)

Romanticism passed its heyday in Europe in the mid-nineteenth century, but in America it still remains an unfinished chapter. It took on a unique Transcendentalist American form with Emerson and Thoreau, and its libertarian flavor still survives in a wide range of American movements, including hippies, environmentalists, and anti-federalist militias. Many of these groups apparently looked at the Indigenous peoples of the Americas and decided that they exemplified the idea of nature over culture, one of the multitude of concepts associated with a Romantic worldview (Honderich 2005: 821).

The conflation of environmentalism and Indigenous peoples began in the 1960s as several phenomena began to converge in an unexpected fashion that served to link American Indians and conservation in popular American culture. First, young Native Americans who were struggling for the recognition of treaty rights of Indian tribes started placing emphasis on traditional values, knowledge, and belief systems. The American Indian Movement (AIM), one of the most active of these groups, joined with traditional elders to oppose assimilationist tribal governments. Around the same time the burgeoning environmental movement, stimulated in part by the publication of Rachel Carson's book *Silent Spring* (1962), began using Native American imagery and assumptions about Indigenous belief systems and traditions to argue that Indigenous peoples of North America were "natural environmentalists."

The rebirth of Indian political activism, along with demands for restoration of Indian lands, led many Indian people to insist that one of the most compelling aspects of their rights emerged from their historically strong connections to the land and nature. Synergy between the newly emerging

1. In this chapter I would like to acknowledge extensive discussions on this topic with my Indigenous colleagues Ronald Trosper (Salish-Kootenay) and Corey Welch (Northern Cheyenne). Any errors that exist are, of course, my own.

environmental movement and Indian activists who emphasized the concept of Indians as being proper stewards for tribal lands resulted in the creation of a series of romantic images linking Indians to conservation and environmentalism.

A final factor that brought American Indian worldviews prominently into popular culture was the five hundredth anniversary of Columbus's so-called voyage of discovery to the Americas in 1992. Indigenous American activists took advantage of the publicity surrounding this quincentenary to demand increased respect and better educational opportunities, along with the right to control land and natural resources on tribal lands. This activism was accompanied by increasing claims that American Indians had unique belief systems and knowledge, and that these needed to be considered as part of the growing idea of Indigenous sovereignty (C. Wilkinson 2005).

One response was a predictable backlash from political conservatives (Lehr 1992; Lewis 1994; and the ironically named Redman 1999), which increased in intensity after the election of a Republican-dominated U.S. Congress in 1994, one of whose goals was retraction of environmental legislation. In response to these attacks, a book first published in the 1970's that purported to deal with North American Indian Ecology (Hughes 1996) was republished. This text emphasized a romantic perspective on the topic perhaps more than it should have. For example, Hughes, a historian at the University of Denver, puts the following statement on the first page of his book. "Somehow they had learned a secret that Europe had already lost, and which we seem to have lost now in America—the secret of how to live in harmony with Mother Earth, to use what she offers without hurting her; the secret of receiving gratefully the gifts of the Great Spirit. (Hughes 1996: 1).

On the face of it, there is nothing wrong with the statement made by Hughes; it simply relies too much on romantic cliché and not enough on the hard-won basis of TEK. More to the point, despite using the term "ecology" in his title, Hughes never directly addresses the true nature of ecology and links among species.

As this controversy over land and environmentalism unfolded, an anthropologist at Brown University, Shepard Krech III, published a book entitled *The Ecological Indian* (Krech 1999a; henceforth referred to as *TEI*). Krech's stated purpose was to "rekindle debate about the accuracy of the stereotype of the Ecological Indian" (28), and to discuss the difference between popular myth and historical reality with regard to the ecological knowledge and understanding of American Indians.

What was surprising for a book supposedly targeted at a scholarly audience is that *TEI* became a moderate bestseller and created a groundswell of attention. As a result, reviews of *TEI* appeared in newspapers and non-academic journals and magazines, including the *Atlantic Monthly, New Republic, Nation, New Yorker*, and *New York Times Review of Books*. The journal *New Scientist* did not bother with the objectivity of a review; they simply had Krech summarize his views in their pages (Krech 1999b).

It appears that Krech's goal was to construct a postmodern response to the romantic tradition within the American environmentalist movement as exemplified by Hughes (1996). A number of individuals associated with the "wise-use movement," which purports to be a rational response to romantic excess, recognized in Krech's work a weapon that they could use to bludgeon both environmentalists and Indigenous activists. Many reviews in the popular press were written by political conservatives, or at least by anti-environmentalists or defenders of the benefits of colonialism, all of whom enthusiastically endorsed Krech's work, which they perceived as providing academic support for their political views (e.g., Lemman 1999; Redman 1999).

Despite the popularity of *TEI* and the use of the term *ecological* in the title, only one review of this book has been written by an actual working ecologist—Cherokee forest ecologist Thom Alcoze reviewed *TEI* in the journal *Ethics, Place and the Environment* (Alcoze 2001). The next closest approximation I was able to find was the *New Republic* review by Stanford University's environmental historian Richard White. Although not trained as a scientist, White understands both the principles of ecology and the history of Indigenous America well enough to evaluate the ecological content employed in *The Ecological Indian* and was not impressed with Krech's accomplishment.

The social sciences and humanities are the two academic fields into which American Indian studies programs are traditionally ghettoized (see Chapter 10). This suggests that the American literary and academic establishment thinks of Indigenous peoples only in terms of social issues, even when they are pointedly characterized as "*Ecological* Indians." Apparently because Krech is an anthropologist rather than an ecologist, most academic reviews of *The Ecological Indian* come from the social sciences and humanities rather than from the natural sciences. The problem with this approach can be illustrated by examining the review by Mark S. Fleisher in the *New York Times Book Review*. Fleisher is identified as the author of *Warehousing Violence* and *Beggars and Thieves: Lives of Urban Street Criminals*, which certainly qualifies Fleisher to discuss stereotypes, but he appears to know very little about ecology or natural science. As a result, Fleisher accepts Krech's statements on ecology at face value. Scholars from the social sciences and humanities seem unable to adequately address the scientific issues raised in *TEI* and have been misled by the excessive number of literature citations to believe that it is well documented. In addition, any mistakes Krech might have made in his statements about Indigenous worldviews remained inadequately addressed, because most reviewers have been non-Native.

Suffice it to say that *TEI* is not the carefully constructed, thoroughly documented synthesis described by many conservative reviewers. Ironically it is also not the speculative propaganda described by *Indian Country Today* (11–18 October 1999) and in an unpublished review by Vine Deloria

Jr. As the Penobscot scholar Darren Ranco points out, "Without the introduction and epilogue I believe that *The Ecological Indian* is an overview of scholarship with little political or intellectual importance" (Ranco 2007: 33). Following Ranco, I suggest that with the exceptions of its opening and closing chapters, *TEI* is a book that includes a wide range of information that is highly variable in quality and does not support any particular point of view.

From my perspective as an ecologist, where *TEI* is weakest is in its understanding of ecology, which Krech chose to emphasize by making it part of the title and referring to it in the text. Like Hughes (1996) before him and a host of scholars that followed him and developed themes inherent in *TEI* (Kay and Simmons 2002; Harkin and Lewis 2007), Krech seems to treat ecology not as a science, but instead as a set of social attitudes. Ranco alludes to this, but not being an ecologist by training, he does not directly make this point.

Reviews of *TEI* tend to fall into three categories: (1) thoughtful reviews by academics outside the natural sciences who credit Krech with having raised an interesting issue, but almost all of whom point out that he has not developed his arguments well; (2) Indian writers or scholars and social activists, such as Deloria, who seem to reject Krech's arguments on political grounds; and (3) political conservatives, including a few academics, who claim Krech has finally laid to rest the cliché of Indians as natural ecologists. Krech does not appear to be a political conservative himself, and he is certainly not anti-environmentalist in his personal views. Krech co-edited the *Encyclopedia of World Environmental History* (Krech and Merchant 2004). Krech appears to be the ultimate postmodern critic—that is, he deconstructs everything yet presents no clear resolution or solutions.

REFIGURING THE ECOLOGICAL INDIAN

The interest and controversy stirred up by *TEI* led the American Heritage Center at the University of Wyoming to organize a conference, which they titled "Refiguring the Ecological Indian," as its tenth annual symposium in April 2002, with Krech as a featured speaker. This symposium was the first venue in which scientists of Native American ancestry were able to respond to Krech and critique the ideas described in *TEI*. Indigenous speakers at this symposium were almost unanimous in their criticism of *TEI*. More to the point, a number of scientists of Euro-American ancestry were also critical of *TEI* for ignoring topics that appeared to counter some of Krech's premises. Telling examples included the fact that agriculture at Zuni Pueblo is at least 2,000 years old, and the condition of soil in Zuni agricultural fields is superior to that of surrounding areas because of the use of traditional agriculture (Havener and Williams, conference abstracts). Similarly, it was demonstrated that despite Krech's apparent criticism of tribal use of fire, traditional tribal burning practices were important in shaping ecosystems (Johnson and Stark, conference abstracts), a point also made by Alcoze (2001).

As one of the few academic ecologists who studies Indigenous TEK, I was invited to this symposium to present one of the final summary papers and to participate in the closing panel discussion with Charlotte Black Elk and Curly Bear Wagner. Subsequently, a volume was produced from this symposium (Harkin and Lewis 2007), in which all the Indigenous participants are mentioned in the Foreword, but only one of the Indigenous scholars who participated in the symposium was invited to submit (Ranco 2007). Thus we have a discussion of Ecological Indians in which only one Indigenous scholar is given a voice, which illustrates the point made in the title of this chapter. As Ranco states, "Asking or assessing one culture to meet the standards of another culture's stereotypes of them seems a dubious project" (Ranco 2007: 34).

Ranco points out (above) that the main reasons *TEI* attracted attention in the popular media are its Introduction and Epilogue, which are particularly misleading and colonialist in their approach (Ranco 2007). In the symposium volume the Canadian anthropologist Harvey Feit does a masterful job of deconstructing Krech's chapter on the overtrapping of beaver in Northern Canada, showing that Krech misrepresents both history and anthropological data (Feit 2007). I will not repeat the points made by Feit and Ranco at length, but instead will review other aspects of *TEI*, highlighting where I feel that Krech is reasonable in his statements, but also where I think Krech goes wrong in both his understanding of ecology and of the attitudes and perceptions of Indigenous people.

Krech opens *TEI* by invoking the image of Iron Eyes Cody as the Crying Indian, an image used in an anti-pollution advertising campaign in the early 1970s. The advertising campaign was designed to give the impression that Indian people would be moved to tears over what has happened to their sacred land. The crucial thing to realize, however, is that this campaign had nothing to do with any real Indigenous people; it was simply a highly effective bit of advertising designed and carried out entirely by people of Euro-American ancestry.

The Crying Indian image is loaded with irony that Krech misses completely. He identifies Cody as a Cherokee actor, when in fact Cody was an Italian, born Espera DeCorti in Gueydan, Louisiana, the son of Italian immigrants Antonio DeCorti and Francesca Salpietra ("Iron Eyes Cody"). Cody (DeCorti) was hired for the anti-pollution advertisement campaign because he was known for playing Indian roles in more than fifty Hollywood films. DeCorti took the name Cody not to indicate any Indian ancestry but in homage to Buffalo Bill Cody and his Wild West Show. Ironically these Wild West Shows appear to be where most of Euro-American popular culture got their original images of Native peoples, and why so many people of European heritage to this day seem to believe that all Indigenous

Americans come from the Great Plains—they wear long war bonnets, and carry lances and associated horse culture paraphernalia.

Krech's decision to hang the major metaphor of his book on a Euro-American advertising campaign rather than an actual Indigenous experience symbolizes the basic approach followed throughout *TEI*. Rather than checking the veracity of his information with Indian people, instead Krech provides a long list of unfiltered literature citations that confuse the reader and information that has apparently undergone little fact checking, such as his assumption that Iron Eyes Cody was an Indian.

The attention *TEI* garnered in the popular press seemed to result primarily because it supposedly calls into question the idea that Indians behaved in what Krech refers to as an "ecological manner." As I mentioned above, however, Krech never discusses ecology as a science, but instead deals almost exclusively with the misunderstandings of Euro-American popular culture concerning ecology, including confusion of the relationship between the science of ecology and "environmentalism." The basic difference is that *ecology* as a science deals with all species, including humans, and looks at the interactions between the living (biotic) and physical (abiotic) worlds, whereas *environmentalism* deals almost exclusively with human impacts upon their own living conditions. This difference between worldviews, one that acknowledges humans as simply one species among many (ecology) and a human-centered worldview (environmentalism), mirrors the themes presented in this book. Krech's conflation of these two worldviews generates the major source of confusion in *TEI* and in books designed to follow up on its popularity.

Krech defines an "Ecological Indian" as follows: "the dominant image is of the Indian in nature who understands the systematic consequences of his actions, feels deep sympathy with all living forms, and takes steps to conserve so that the earth's harmonies are never imbalanced and resources never in doubt" (Krech 1999a: 21). The first part of this definition simply refers to the nature of Indigenous knowledge or TEK, which I have discussed in detail in previous chapters. The second part of the definition is a statement about worldview and values, which I have also addressed in previous chapters, although I would replace the phrase *feels deep sympathy*, which is emotional and subjective and indicates the romantic perspective from which Krech writes, with the phrase *recognizes connections and relatedness*, which carries less romantic baggage and is more true to the understanding I have of the way Indigenous people view other life forms.

The third part of this definition involves two aspects that are somewhat contradictory; I have argued that Indigenous people developed rituals and ceremonies specifically designed to minimize chances of resource collapse, yet I question that there is any general assumption among Indigenous people that "the earth's harmonies" should be maintained. As indicated above, Hughes (1996) also employs the romanticized concept of "harmony," to describe the relationship between Indigenous people and the land and another romantic. Harkin (2007), states that, "Salmon were thought to come from a place of infinite wealth and to represent a steady supply of food." As I make abundantly clear in this book, Indian people had hard-won knowledge of the ecosystems of which they were a part, and their spiritual practices were designed to cope with the fact that resources were always variable and hence "in doubt."

A more appropriate title for Krech's book would have been *The Environmentalist Indian*. The topics he addresses represent concerns of "conservationists," "resource managers," and "environmentalists," rather than those of scientific ecologists, and his themes are firmly within the realm of environmentalism rather than ecology. The major problem is that Krech's criticisms are directed at a white romantic construction of Indians rather than any Indigenous reality that has ever existed. Neither "noble Indians" as imagined by the conservation movement and described by Krech, nor the ecologically destructive communities conjured up from his "evidence" by conservative commentators have any meaningful basis in reality. Both are products of European imagination—Indigenous cultural traditions are based on ties to specific places or areas, and traditional tribal peoples recognize themselves as simply one species among many, not inherently superior to any other species but dependent on the other species. As stated by the Lakota Luther Sanding Bear, "For the animal and bird world there existed a brotherly feeling that kept the Lakota safe among them" (Standing Bear 1978: 193).

Krech's ideas concerning whether or not Indigenous Americans functioned according to some Euro-American construct of an ideal environmentalist fails to recognize the reality of Indigenous lives. It is important to recognize that Indigenous people are not Romantics in the sense of the European Romantic movement. Europeans and their American descendants have tried to force this image upon them and some Indian people exploit it for economic gain (e.g., Storm 1972; McGaa 1995; for a more thorough review of this topic see Rose 1992). Europeans and Euro-Americans also created the idea of the American West as an Edenic paradise filled with vast numbers of animals, but it is very unlikely that any of the Indigenous peoples who lived there saw it this way. To them it was a place to live, where if you were smart enough and skilled enough the land would allow you to live a decent life upon it. Completely ignored in *TEI* is the fact that Indians were, and many continue to be to this day, practical people who depended on their knowledge of the natural world to survive (e.g., Kawagley 1995).

THREE LEVELS OF ERROR

There seem to be three levels of error in both Krech's thinking and interpretations of *TEI* that weave around each other in confusing ways that result in odd conflations. First and most important, Krech is incorrect in equating

Indigenous beliefs and environmentalist political perspectives, even though this mistake is often made by environmentalists themselves, as well as other adherents to the modern romantic philosophical tradition. This topic is an important subtheme of my book, and I explore it in detail in Chapter 11, where I argue that environmentalists or conservationists are, in fact, very different from Indigenous people in their approach to nature.

Second, as mentioned above, when discussing the reviews of *TEI*, political conservatives who view Krech favorably employ arguments from *TEI* to criticize environmentalists, an approach that can be seen in Nicolas Lemman's review in the *New Yorker* entitled "Buffaloed" (13 September 1999). Lemman's politics are hard to discern, except that over the years he seems to have been consistently anti-environmentalist in his writing. It is curious why the *New Yorker* chose a biased reviewer who lacks the scientific credentials to comment on these matters (Lemman is dean of the Columbia School of Journalism and not a scientist of any kind). This attitude is revealed in his review, which is also nominally a review of Calvin Martin's (1999) *The Way of the Human Being*. However, of the nearly three full pages of this review, only one paragraph is devoted to Martin, who Lemman quickly dismisses, with the remainder being a laudatory discussion of the "cautious, unbiased evidence examiner" [*sic*] he perceives Krech to be. The fact that Lemman juxtaposes Martin and Krech in a review like this is telling. One of Krech's earlier works was an edited volume critical of Martin's *Keepers of the Game* (1978). *Keepers of the Game* argues that the devastating impacts of introduced disease on Native American populations and cultures led them to break long-standing relationships with their nonhuman relatives, which may have led to drastic declines in local populations of beaver and other fur-bearing mammals. Martin's major works: *The Way of the Human Being* and *Keepers of the Game*, provide a very different view of Indigenous worldviews than those presented in *TEI*.

According to some reviewers, Krech's vague comments can be interpreted as a refutation of the idea that Indigenous peoples adhered to Euro-American environmentalist principles. This point is largely irrelevant, because Indigenous peoples did not in fact adhere to such principles, which were not part of their philosophical tradition. Antipathy directed against anti-development attitudes shown by environmentalists has been used to generate attacks, which then focus on Indigenous peoples and their belief systems (e.g., Lehr 1992; M. Lewis 1994; Lemman 1999; Redman 1999), even though none of the participants, regardless of political orientation, has presented an accurate portrayal of how Indigenous people think or feel about the natural world.

This leads to the third level of error in Krech's approach, the failure to understand the basis of Indigenous knowledge or TEK, which is encoded in stories that present ideas that function as metaphors or models based on solid empirical observation of natural phenomena. As human beings dependent on a variable natural world, Indigenous peoples developed cultural traditions that allowed them to respond to the changing environmental conditions that they experienced. Indigenous people functioned ecologically as predators. Harkin (2007) seems to have mixed feelings on this point—despite arguing that Indigenous peoples functioned as predators, he contends that "The existential fact that other living beings must die so that humans can continue is a source of contradiction elaborated by many religions and ideologies . . . The famous metaphorical elaborations of (Indigenous) people . . . seems a particularly rich symbolic elaboration of this existential fact. Such a philosophy of the world cannot, and never could, provide a road map of resource management . . . Rather than a road map, they provide poetics; rather than technology; the consolations of philosophy" (Harkin 2007: 229). The number of odd assumptions in this statement makes it hard to deconstruct. Basically Harkin seems to argue that failed theoretical models of resource management are superior to stories that emphasize connections and links between humans and nonhumans as a possible strategy of conservation. I have already pointed out that Harkin regards the Northwest coastal environment as so rich that "Salmon came from a place of infinite wealth" and represented "a steady supply of food." Harkin's (and Krech's) failure to recognize the variable nature of environments and the corresponding unpredictability of food supplies, shows how they are locked into the ecological thinking of the 1950's.

Like all predators, regardless of species, Indigenous hunting and fishing societies experienced periodic fluctuations in prey availability. There existed a constant risk of overexploiting local prey populations, especially if those populations were experiencing poor environmental conditions. Such consequences were illustrated in *Keepers of the Game* (Martin 1978), where it is argued that the impact of disease on ungulate populations would also have impacted the populations of predators that depended upon these ungulates as a primary food source.

The potential impact of disease introduced into indigenous American populations of ungulates by the arrival of European livestock goes completely undiscussed by any scholar but Martin (1978). As an example, the volume edited by Kay and Simmons (2002) argues that the large ungulate populations experienced by European "explorers" were the result of the decimation of Indigenous human populations who served as the "primary predator" on these ungulates (with no apparent consideration of wolves, who are but a footnote in this book), without a single thought to the fact that animal derived diseases might also have devastated the wild ungulate populations, thus inflicting a double calamity upon the Indigenous peoples. In an attempt to reduce the chances of such shortages occurring, Indigenous cultures developed belief systems and ceremonies that encouraged behavior by the humans that reduced chances of overexploitation of local resources (Anderson 1996; Raban 1999; Pierotti and Wildcat 2000). As an example, the potlatch ceremonies of the Pacific Northwest were based on the metaphor of alternating cycles of abundance and starvation.

Krech presents evidence that suggests that Indigenous Americans may have at times overexploited some populations, especially of deer or beaver, and his arguments are endorsed and developed by Kay and several participants in his 2002 symposium volume (Alvard 2002; Kay 2002a, 2002b). This is hardly surprising. Given fluctuating environmental conditions combined with hunting pressure, occasional reductions in animal populations are inevitable. Kay and Krech (and Harkin and Alvard) show no apparent understanding of Indigenous belief systems that were specifically designed to minimize such events. In fact, Krech parodies these belief systems. He argues that one reason hunters killed all survivors of a buffalo jump was because, to quote Krech (1999b), "they believed that, as animate beings, escaping buffaloes would tell other buffaloes about the trap awaiting them," and they did not want the survivors to escape and teach other bison to avoid human hunters. From the point of view of scientific behavior and ecology, this is not necessarily a myth. It is well established that prey quickly learn to avoid predators, and this knowledge is almost certainly passed quickly among social groups of prey animals. In fact, this point is argued by several scholars who employ a perspective similar to that used by Krech (Berger et al. 2001; Kay 2002a; Harkin 2007; chapter 10 in Kruuk 2002).

It is essential to keep in mind that prior to the acquisition of horses, Indians hunted bison using spears and bows and arrows, which means they had to get close to large and potentially dangerous prey to make a kill. Once they had horses to help pursue and select individual bison to make kills, Indigenous hunters showed less concern about the cultural traditions of bison, but they were also much more selective in the way they made kills; as a result, potential "waste" was greatly reduced (Isenberg 2000). The rifles employed by Euro-American bison hunters rendered it unnecessary to worry about bison behavior since animals could be shot from a great distance without even realizing that they were at risk. Ironically in the long run, this alleged "Indian myth" may be a more accurate representation of the response of social prey to predators than the idea promulgated by Paul Martin and his acolytes that large mammals continued to be so naïve as to be susceptible to repeated attacks by human predators, which led to their extinction (P. Martin 2005; Kay 2002b; see Chapter 9 for more discussion of this point).

Also troubling is Krech's misunderstanding of Indian beliefs in his discussion of population distributions, dispersal, and movement patterns as well as rituals designed to assuage prey populations and keepers of the game. Krech points out that some tribes (Cheyenne, Arapaho) had a tradition that allegedly believed that when hunters were unable to find bison this was because the bison had retreated to an underground prairie (*TEI*: 145–147), and contends that this demonstrates that "[Indian] ecological systems embraced components like underground prairies, which were absent from the ecological systems of Western tradition" (*TEI*: 212). I watched Krech employ this argument in defense of his ideas at the American Heritage

Symposium mentioned above. Krech is taking literally something that is meant to be taken as a metaphor. In plains tribes, going below ground is a metaphor for moving beyond the horizon, so such stories clearly refer to migration, rather than to literal underground lands.

Krech's failure to understand the scientific study of ecology and the history of this discipline is obvious. To begin with, the Western tradition contemporaneous with these ideas from Native American tradition predates the development of ecology as a science. Ecology did not arise conceptually as a part of Western science until the late nineteenth century, after Darwin developed his ideas about natural selection and the role of the environment in thinning out natural populations (Kingsland 1991, 2005). *Ecology* was first used as a term in America in the 1890s and did not exist as a formal, rigorous data-based science until the 1920s when the Chicago School developed theoretical models of animal populations (Kingsland 1985, 1991, 2005). As I pointed out at the Laramie symposium during my presentation, the ideas expressed by Okute (quoted in Chapter 4) are not functional equivalents of the ideas of the ecology of the late nineteenth century (during Okute's lifetime), but rather those of contemporary evolutionary ecology developed in the 1980's and 1990s.

As a model, the metaphor of underground populations is not really that different than concepts that exist "from the ecological systems of contemporary Western tradition" (*TEI*: 147). The idea of large reserves of highly mobile animals capable of recolonizing empty habitats is similar to the modern ecological concept of animal metapopulations (Nee 2007). A metapopulation is assumed to consist of local "sinks and sources" in which some areas contain healthy reproducing populations that produce excess individuals (called source populations) that can be forced by density-dependent processes to move into population "sinks" where mortality exceeds recruitment (Hanski 1999; Nee 2007).

It is important to emphasize that the idea of sources and sinks is also a metaphor, albeit one from the Western scientific tradition. Another interesting thing about source sink dynamics is that Krech and his acolytes never mention such recent developments in ecological thought. For example, several authors reviewing populations of bison point out that these animals were most abundant in areas that were not controlled by any one people, which were often zones where conflict occurred, so animal populations were less intensively hunted, whereas populations in areas controlled by one people probably experienced heavy levels of exploitation (Flores 2007; Kay 2002a, 2002b). What goes completely unmentioned is that a mixture of habitats, in some of which animal populations reproduce enough to export individuals into areas where reproduction is more chancy or may not even occur, is a textbook examples of source and sink dynamics (Nee 2007), which suggests that such concepts may have been understood by Indigenous hunters, even though they did not become part of ecological discourse until the 1980's.

This notion of source and sink populations is important. If a population of bison fell to low levels, or even went locally extinct, this could force local human groups that depended on it as a food source to move into areas of potential conflict with other peoples to hunt (Flores 2007). In fact, extinctions of once locally abundant populations of large game, such as elk and deer, did force abandonment of European settlements in western and southern Kansas and northern Oklahoma that depended on these populations to supplement their food supplies when their agricultural crops failed during the latter half of the nineteenth century (Frank Cross, Kansas Museum of Natural History, personal communication).

One of Krech's most compelling examples from the perspective of a number of reviewers is his discussion of buffalo jumps, where bison were driven off of a cliff edge or into a deep ravine. For example, the sole illustration accompanying Lemann's review is of Ralph Steadman's cartoon bison flailing helplessly in the air after being driven off a cliff. In a similar vein, the sidebar accompanying the review of *TEI* in *Discovery* magazine states in bold colored letters: "Some groups killed hundreds of beasts at a time, chasing them over cliffs" (Richardson 1999). Krech argues extensively that these were sites of great waste, yet he never presents any evidence that Indian hunters had negative impacts on bison populations. His basic argument can be summed up briefly: Indians were wasteful of bison at times; therefore they may have had some (undefined) role in the destruction of the great buffalo herds by driving them over cliffs, even though this practice had ceased two centuries before the near extermination of the great herds. Krech's discussion of buffalo jumps, however, is naïve . Buffalo jumps were used as a means of taking bison prior to the arrival of horses on the plains. It must be recognized that it is difficult to control the number of animals being chased and killed under such circumstances.

It may well be true that waste occurred, if you define "waste" solely in terms of human use. Given the number of other predators (wolves, bears, eagles, ravens, crows, magpies, etc.) with which Indigenous human hunters shared the West, I find it unlikely that much meat went uneaten (Welch 1986; Schlesier 1987), at least until Europeans arrived and worked hard to exterminate these other predators along with bison and Indians (Isenberg 2000). During much of this time it is likely that humans hunted cooperatively with wolves (Schlesier 1987; Chapter 3), so even if humans did not use all of the meat it was not "wasted" from the Indigenous perspective of all things being connected and related.

Krech admits that there are problems of interpretation with regard to some of these kill sites, especially in circumstances where a jump or kill site may have been used multiple times over hundreds, perhaps thousands, of years, and little can be said about any single episode. The example that Krech chooses to highlight is a "kill site" in southern Colorado, identified as Olsen-Chubbuck (*TEI*: 144). In this location hunters supposedly "drove a herd of bison over the bank of a dry gulch." This site is over

8,000 years old, and Krech never makes clear how it was established that, given its antiquity, it was identified as "a single episode." Krech focuses his detailed discussion and his arguments on this example, as if this were typical of all buffalo jumps. Krech later acknowledges that most accounts by Indians emphasize that after the butchering of bison was finished, sites were cleaned out, and everything was taken. Therefore it is unclear how Olsen-Chubbuck can be clearly identified as the result of a deliberate drive, rather than the scavenging of a natural event that could have resulted from other causes, such as a prairie fire or severe thunderstorm accompanied by tornadic activity.

Finally, if bison populations remained healthy there was little pressure to behave conservatively. As Richard White (2000) points out in his review of *TEI*, "Conservation is meaningful only when the harvest of a resource exceeds the reproduction of that resource. If hunted populations remain abundant, then 'waste' becomes a meaningless vice." It has been argued by some scholars that some bison populations may have been exterminated at the local level by heavy hunting pressure (e.g., Flores 2007), which fails to recognize or discuss the source sink concept discussed above.

The way Krech describes Indian hunting seems to assume that nonhuman predators never kill more than is necessary or become specialized in eating only parts of prey when prey are abundant. This is simply not true. Wolves have been known on occasion to kill more deer than they can eat in a short time. Bears, eagles, and gulls feeding during salmon runs are very selective in what parts of the fish are eaten. Many nonhuman predators feed on only the most tender or nutritious parts when food is abundant (Kruuk 2002). According to the rather simplistic criteria set out by Krech, (also by Alvard 2002; Kay 2002a, 2002b, and Flores 2007) none of these nonhuman species would be considered "ecological" in their behavior.

MIXED MESSAGES AND CONFLICTING DATA

One major difficulty in commenting on or critiquing *TEI* is that it can be difficult to distinguish what Krech actually says from what his reviewers often credit him with saying. For example, when discussing bison kills, Krech seems to want to have it both ways. First, he argues that Indians were wasteful, yet later he acknowledges that according to Indians' own accounts, they tried to minimize waste.

A more serious problem can be seen in the way in which reviewers have added to the confusion. In his *New Yorker* review, Lemman (1999) makes the claim that "over the years, [Paul] Martin's critics (of the Pleistocene Overkill hypothesis) have come back with the theory that these species were killed off by asteroid showers and climate change, not by Indians" (98). Paul Martin is a faculty member in geosciences at the University of Arizona, who published a series of papers and books arguing that early modern humans were responsible for the extinction of many species of large mammals and birds, particularly in North America and Australia (summarized in Martin 2005). North America and Australia are also the two places where Indigenous peoples have consistently been making the most effective and insistent land claims, so there is considerable political motivation to portray these Indigenous peoples as "wasteful" and "destructive," hence "non-ecological." It is unclear as to whether Lemman's idea is derived from Krech, the scientific literature, or Lemman's imagination, but this does suggest some interesting issues concerning fact checking by *New Yorker* and journalistic training at Columbia. Lemman seems not to understand the difference between large Pleistocene mammals and dinosaurs; the latter may well have been driven to extinction by a single large asteroid that collided with the Earth some 75 to 80 million years ago. In any case I can find no comment in Krech alluding to asteroid showers (whatever these are; the idea of showers usually relates to meteorites). In contrast, it has been quite convincingly argued in the scientific literature that climate change was responsible for the extinction of North American megafauna and strongly implicated in extinctions of European megafauna as well (Guthrie 1990; Grayson 1991; Grayson and Meltzer 2002, 2003; Stuart et al. 2004; C. Mann 2005).

Such confusion about the message contained in *TEI* results largely because Krech often appears unable to make up his mind about the ecological impact of Indigenous peoples. Krech emphasizes the idea that Indigenous peoples had major impacts on their environments and makes the statement that Europeans arrived in a "widowed" rather than a "virgin" land (99). Later in the same chapter Krech argues that Indigenous populations were never large enough to have had any significant impact. This argument runs contrary to recent arguments advanced by Charles Mann, which support the idea that Indigenous peoples had major impact (see also Kay and Simmons 2002); however, Mann emphasizes that these impacts resulted because human populations of the Western Hemisphere were much larger than has traditionally been thought (C. Mann 2002, 2005).

There is considerable controversy over the size and distribution of pre-Columbian human populations throughout the Americas. It is generally argued that based on the size of the population at the time of European settlement along the Atlantic Coast in the seventeenth century, the total Indigenous human population of North America was around two million. Later, as Europeans moved across the North American continent they continued to estimate populations of Indigenous peoples. No matter what method was used, the results always seemed to indicate a roughly similar total population in the vicinity of two million. Thus it was concluded that introduction of diseases such as smallpox, typhus, influenza, diphtheria, cholera, and measles by Europeans had no measurable impact on Indigenous populations (C. Martin 1978; Krech 1999a).

The idea that Indian populations were unaffected by introduction of disease, which Krech appears to support, is a self-serving and inaccurate way of viewing the impact of epidemic disease at the social, cultural, and ecological level. Discussing estimates of Indian populations, Mann points out that, in fact, "this picture of Indian life is wrong in almost every respect. Indians were here far longer than previously thought . . . and in much greater numbers" (C. Mann 2002: 41). Mann's 2005 book presents considerable data to support this argument, which is also endorsed by most of the scholars in Kay and Simmons (2002). Krech waffles considerably in his arguments concerning total Indigenous population size, and appears to be uncomfortable with the idea of a stable and unchanging Indigenous North American population around two million. Krech finally opts for a middle ground of a total pre-Columbian contact population size between four and seven million. This estimate is still very low compared to evidence presented by Mann (2005). Krech's figure of four to seven million does not seem to be based on real data. Like many of Krech's conclusions, it simply seems to be a contrived compromise that results from his attempt to be all things to all readers.

The point raised by Mann (2002, 2005) is that impacts of the First Americans are everywhere and are still obvious to those who take the time to look. Colonial Europeans simply were not equipped to understand the difference between managed landscapes and "pristine" landscapes because they really had little idea of what "pristine environments" would look like (C. Mann 2005). It is perhaps noteworthy that Hughes (1996) makes naïve assumptions concerning the impact of Indigenous peoples, even though he is endorsing the idea of "environmental Indians." Europe in the seventeenth century was so disturbed ecologically that even environments profoundly impacted by Indigenous people looked pristine compared to the heavily impacted environments of Western Europe.

Some scholars seem to follow the tradition established by Krech, while at the same time acknowledging the fact that Indigenous Americans were much more numerous than has been generally recognized (e.g., Kay and Simmons 2002). This allows them to avoid the confused argument that traps Krech. The basic perspective in this newer approach is that Indigenous Americans were so numerous that they must have had significant impacts upon nonhuman populations that they exploited for food and shelter. This is certainly credible, although some investigators extend this argument to support of the idea of Pleistocene overkill (Kay 2002b), developing their own concept of what they refer to as "Aboriginal Overkill" (Alvard 2002; Kay 2002a).

The problem with linking Aboriginal hunting over the last 2,000 years with Pleistocene overkill is that, although it is undeniable that Indigenous populations reduced numbers of prey species, especially on a local level, there is no evidence of extinction in any of these species. As one participant in Kay and Simmons puts it, "Pleistocene overkill requires that the tiny founding populations of New World peoples *swiftly hunted to extinction 35 separate genera of large mammals*, while the substantial harvest pressure and *severe impacts of the enormous late Holocene human populations required millennia and failed to cause the extinction of a single species*" (Broughton 2002: 67, emphasis added). The one question rarely mentioned by Euro-American investigators who support the idea of "Aboriginal Overkill" (Martin, Krech, Harkin, Alvard, Kay, etc.) is that "If they were so successful as hunters that they wiped out megafauna as they encountered them across North America, why as their numbers increased and they became well established (Alvard 2002: Kay 2002a) were there no further extinctions?"

Given the blitzkrieg metaphor employed by Martin, cited by Krech, and refined by Kay and Alvard, we should expect continual extinctions, yet it was not until Europeans arrived with their combination of firearms, disease, and hostility towards nature (Guns, Germs and Fear?) did any other species go extinct, at which point a rolling wave of extinctions took place as Europeans arrived, starting in the eighteenth century with Steller's sea cow and the Great Auk, continuing with the Atlantic gray whale and Carolina Parakeets, extending through Passenger Pigeons and eastern bison, and culminating in the 1880's with the extinction of Northern fur seal populations outside of Alaska and near extinction of plains bison, California sea otters, and Northern elephant seals, combined with extermination of wolves and grizzly bears from almost all of their range within the continental United States.

One individual follows Martin's lead by arguing that, based on simulations of ungulate populations hunted by wolves in Alaska, a small amount of added human predation on the ungulates can lead to a rapid decline (but not extinction) of ungulates (Kay 2002a), which supports the idea of Indigenous extinctions of megafauna. The problem with this approach is that, as indicated above, this is a simulation model, which has numerous simplifying simulations, including the assumption that only a single alternate prey exists for either wolves or humans. This approach is similar to those I described in Chapters 1 and 4, of scientists believing their models to be accurate representations of reality, rather than the metaphoric equivalent of stories.

In fact, the major problem with the model cited by Kay (2002a, 2002b), is that in the original paper from which this simulation is derived (Walters et al. 1981), it is clear that this system was initially perturbed by a 75 percent decline in Dall sheep populations that occurred in the 1930's when Euro-Americans moved into interior Alaska (Figure 3 in Walters et al. 1981). The simulation upon which Kay relies for his explanation of Pleistocene overkill assumes that the sheep population is stable at roughly 20 percent of what it had been in the 1920's, when it is clear that this was a seriously impacted population. Interestingly, the Kluane people in the Yukon argued that a similar decline, caused by hunters of European ancestry, affected Dall sheep populations in their area, which is why they felt that hunts for this species should be curtailed (Nadasdy 2003). To summarize, Indigenous peoples are excoriated for having "possible impacts" on some prey populations in an effort to compare them directly to European invaders who definitely caused extinctions and near extinctions wherever they went.

A final point needs to be made here. There is a school of thought that appears to start with Brightman (1993) and continues through Krech (1999a) and Harkin (2007) that indigenous cultures had no understanding of conservation until they were taught these principles by Europeans. This ties directly to the Overkill hypotheses, both Pleistocene (Martin 2005) and Aboriginal (Kay 2002a). Along with the question of why there have been no extinctions of mammals or birds since the Pleistocene raised by Broughton (2002), this line of argument asks us to accept the idea that all of the spiritual traditions and ceremonies related to links to the natural world and placation of prey species (Amoss 1987; Marshall 1995; Harrod 2000; Martin 2000; Kidwell et al. 2001; Atleo 2004; Alessa 2009) have been created from whole cloth since first contact by Europeans. This line of thought is so naïve and insulting at so many levels that it is difficult to conceive of a more colonizing way of looking at the world.

ARE THERE "NATURAL" PLANT COMMUNITIES? THE ROLE OF FIRE

It is obvious to contemporary ecologists that many habitats in North America are dependent upon fire (sometimes called "fire climax" communities). The use of fire by Indigenous peoples is one clear and unequivocal example of an Indigenous impact that shaped ecological communities in the Americas (M. Lewis 1989; Boyd 1999; Barsh 2000; Williams 2002; Kay 2002b; Levy 2005). Krech discusses this phenomenon in a way that confuses the issue. He starts out reasonably by pointing out the illogic of the historical Euro-American "Smoky the Bear" perspective (which was held by Swiss-American conservationist John Muir, among others), that all fire is harmful and destructive. Krech then discusses the many ways in which Indians employed fire, such as to clear underbrush, create better grazing conditions for ungulates or better habitats for food plants. Krech then segues without pause into a discussion of Indian-set fires that may have gotten out of control and caused damage, or even had an outcome contrary to that intended. Missing from Krech's chapter on fire is any discussion of modern fire ecology, where it is generally appreciated that fire is a positive influence in maintaining habitats suitable for hunting, plant harvesting, and so forth (Lewis 1989; Boyd 1999; Williams 2002; Wuerthner 2006).

Krech's claim to understanding Indian people is based on his "fourteen months in 1971 to 1972 among the Gwich'in who lived in . . . Canada's Northwest Territories; I returned for several weeks in the fall of 1984" (Krech 1999a: 9). The Gwich'in people represent a fairly complex example to discuss in this context of Indigenous use of fire, because they exemplify the problems involved in conflicting views of how fire functions in ecological systems (Levy 2005). "Fire suppression has made it increasingly difficult for native Gwich'in people to use their traditional means of subsistence. . . . Before the state government cracked down on the practice . . . the Gwich'in routinely burned along creeks and wetlands to create fresh forage for moose, muskrats, and waterfowl. Now . . . Gwich'in are caught in a conflict between their dependence on fire-fighting jobs . . . and their need for flames in the boreal forest" (Levy 2005: 307).

Re-creation of Indigenous burning patterns in Redwood National Park has led to an increase in native species, a drop in fuel loads, and a reduction of numbers of invasive exotic plants (Levy 2005), which suggests that traditional Indian burning practices generally have positive impacts. One scholar describes it thus, "Generally native people burned parts of the ecosystems in which they lived to promote a diversity of habitats . . . which gave the Indians, as well as animals greater security and stability in their lives" (Williams 2002: 196). This finding also helps to refute the claim made by Harkin (2007) that Indigenous peoples did not have reciprocal relationships with nonhumans. The alteration of habitats that allowed prey species to increase in number and safety clearly sets up a reciprocal relationship, as was argued by Langdon (2007) for North Pacific Salmonids.

In *TEI*, Krech ends up assuming a position similar to those who criticized the practice of controlled burning because fires that were set to remove artificially high concentrations of brush and deadfall around Los Alamos, New Mexico, got out of control. In this case a controlled burn meant to reduce the chances of serious wildfires caused considerable damage when it got out of control because of unexpected winds. A single instance in which a fire gets out of control does not invalidate the practice of controlled burns. Such burns are typically effective in reducing the intensity of wildfires and probably would have reduced the destruction caused by fires in coastal Southern California and in Yellowstone Park in the 1980s. Weather conditions can change, and often do so quickly, and fires sometime get out of control or cause unintended damage. This does not mean that individuals who employ fire are irresponsible or "not ecological." To Indigenous elders in California, "the most important reason for burning is to keep down the brush, to prevent catastrophic fire" (M. K. Anderson, cited in Levy 2005). After Indigenous peoples were removed, either by disease or removal to reservations, the nature of forests changed from "parklands" to "dense and scrubby with impenetrable thickets of vegetation beneath the woodland canopy" (Botkin 1990; Williams 2002). The general consensus is that fire is useful, especially in the manner employed by Indigenous peoples worldwide (Lewis 1989; Boyd 1999; Levy 2005).

Krech's discussion of fire reveals two major points about the attitude displayed in *TEI*: 1) Krech oversimplifies and caricatures both ecology and Indian burning; and 2) it appears that having any noticeable impact makes a people "non-ecological." Following Krech's logic, Gwich'in jobs as firefighters would seem to make them "ecological" (i.e., environmentalist), whereas their traditional burning patterns that increase forage and biodiversity could be considered "non-ecological." Thus, he falls into the trap he presumed to set, declaring that to be truly ecological, Indians could not have had any discernable impact on their environments, a patently absurd premise.

Virtually all species have an impact on their environments, especially on a local or microhabitat level (Lewontin 2001). This has become more appreciated as modern ecologists have come to recognize the importance of what is now referred to as ecosystem engineering (Jones 2005; Wright and Jones 2006). One of the best examples of how nonhuman organisms alter local environmental conditions is the beaver, *Castor canadensis*. This large, dam-building rodent creates major changes in local ecology by damming streams, thus creating wetlands and promoting certain forms of ecological succession. In the American Southwest, erosion led to the creation of arroyos and flash flooding that resulted from removal of beavers from nearby mountains. Restoration of beavers into some areas has reduced both erosion and flash flooding. In addition, the American prairie, which was considered to be the quintessential American habitat, was maintained largely through fires, some natural, some set by Indigenous peoples, as well as by heavy grazing from bison and other large ungulates.

To summarize, "The basis for much of the 'forest health crisis' of today really started with the almost complete cessation of Indian burning during the early 1700's in the East and the 1850's in the West" (Williams 2002: 207). This suggests that Indigenous peoples were both an important part of the ecological communities in which they functioned and that their activities definitely benefitted many of the species with whom they interacted on a regular basis, thus establishing the reciprocal relationship denied by Harkin (2007) in his attempt to support Krech's arguments.

AN EMBARRASSMENT OF RICHES

The chapter on fire exemplifies another major problem in *TEI*. Many of Krech's reviewers mention how impressed they are with his voluminous notes and citations, which make up more than 25 percent of the total text. The very volume of these citations creates its own problem, because this apparent embarrassment of riches makes it very difficult to link many citations with the evidence they are purported to supply and the statements they are supposed to support. Several of Krech's citations occupy more than a page of text and contain more than 30 individual references, yet the endnote number appears to apply to a single summary statement. This structure leads the naïve reader to the assumption that every one of these numerous references says what is included in Krech's statement that accompanies the notation. This seems very unlikely, and in many cases statements made by Krech cannot be linked to specific statements in any of the sources provided. Krech appears to treat all evidence as equally valid and does not analyze the evidence he presents to indicate which sources are authoritative, and which accounts are questionable, or that secondary and even tertiary sources are not as well documented as primary sources. This is a serious issue when evaluating scholarly works. As a counterexample, Mann (2005) carefully documents his statements and arguments and typically provides only one or two citations per footnote, as do the authors in Kay and Simmons (2002), which makes their information much easier to verify.

Krech is also given to vague statements. In the chapter on fire (*TEI*: 107) we are told, "Throughout the East, Indians cleared hundreds, and at times thousands of acres for their crops" and "In Virginia, *mention was made* of thousands of acres of cleared land under cultivation" (emphasis added). No reference of any kind is associated with either of these statements. Statements of a factual nature should have one or at most a few citations that directly support the statement. When the use of clauses such as "mention was made" are employed, the quality of information is called into doubt. What does this clause and the lack of citation mean? That it is based on hearsay?

In addition, I find it interesting and instructive that despite the huge volume of citations, there are some significant omissions. Indigenous writers are largely ignored in *TEI*. There is no mention of Luther Standing Bear, and Vine Deloria Jr. and Charles Eastman (Ohiyesa) are only mentioned in passing. In addition, significant work on Indigenous perceptions and cultural traditions by non-Indians are also ignored. There is no mention of the research and writings of either Calvin Martin or Robin Ridington, both of whom emphasize the nature of Indigenous peoples as predators, and both have spent considerable time living and working in Indigenous communities.

The most important omission is a book that deals with a topic similar to *TEI* but has the advantage of being carefully argued and thoroughly documented: Eugene Anderson's *Ecologies of the Heart* (1996). Anderson's book, published by a prestigious academic press, demonstrates that Indigenous peoples around the world have developed empirically based spiritual traditions that link them to their local environments and the variable conditions found in such environments. In particular, Anderson discusses how belief systems are linked to fluctuations in resource abundance. This and other works clearly related to Krech's topic are simply not discussed in *TEI*. Sadly, I suspect that their inclusion would demonstrate the logical weaknesses of *The Ecological Indian*.

Instead of citing Indigenous scholars and writers, and Euro-American writers such as Anderson, Martin, and Ridington, Krech cites purveyors of misleading noble Indian images such as Ed McGaa (Eagle Man) and Sun Bear and opens and closes his book with Iron Eyes Cody. The reason seems to be obvious. Eagle Man and Sun Bear provide useful examples of overly romanticized "Environmental Indian" style thinking, just as Iron Eyes Cody portrays a fictional character that supports the stereotypical view of Indians as environmentalists.

The Ecological Indian fails, primarily as a treatise on ecology, and secondarily as a treatise on Indigenous knowledge and understanding of nature. Thus it is true to neither of the words in its title. To effectively evaluate the ecological aspects of human behavior and cultural traditions it is necessary to consider how the cultural traditions of Indigenous Americans come into being, how these traditions serve to function within ecological systems, and differences between opinion and evidence in evaluating the actual ecological impacts of any given action. Thus, my conclusion is similar to that of Ranco (2007) that aside from the Introduction and the Epilogue that *The Ecological Indian* is not a significant book. If the reader wants to read a more carefully reasoned and thoughtful exposition of similar arguments, I recommend Kay and Simmons (2002). Krech fails to perceive the significant philosophical differences between European and Indigenous traditions that are the theme of this book—that TEK and other Indigenous perspectives are derived from the assumption of being part of the local ecosystem, whereas Western (European) thought derived from the Romantic movement and Baconian/Cartesian mechanistic thinking, clearly places humans as separate from and hierarchically above nonhuman elements of the community. The impacts of Indians were everywhere, but they were qualitatively different than the type of impacts made by Europeans, both in their own land and in the Americas after their arrival (Broughton 2002; Williams 2002).

The major failure of Krech in *TEI* is that he does not comprehend the difference between attempting to set up a worldview that works to minimize harmful impacts upon the nonhuman world with the idea that no impact at all is ever caused. All cultural traditions, both human and nonhuman, have impacts, and sometimes these impacts have adverse effects. The key difference is that Indigenous philosophy is designed to minimize and consider the effects of such impacts. In contrast, the Euro-American perspective seems to treat the nonhuman world as either an opponent or from a romantic perspective as a "treasured loved one who cannot be harmed at all."

This attitude has led to the removal of Indigenous peoples from national parks or wilderness areas on the theory that humans can never be part of the natural world, which exists only in a state without humans (Marshall-Thomas 1994; Dowie 2005). This is harmful from an ecological perspective because by removing humans, we are damaging all the co-evolved responses and relationships that humans have established with nonhumans over the last several thousand years (see Kay 2002b for a more dramatic argument on this topic). *TEI* deals only with whether or not Krech's environmentalist standards are met. As Richard White states in his review, "[Krech's] verdict seems based less on the existence of 'Ecological' thinking than of whether Krech approves of the results" (White 2000).

9 A Critical Comment on Both Western Science and Indigenous Responses to the Western Scientific Tradition

> The reason that people do not have a correct view of nature is not that they are ignorant of this or that fact about the material world, but that they look to the wrong sources in their attempt to understand.
>
> Richard Lewontin (1997)

> A traditional Indian finds himself still experiencing the generalized presence of spiritual forces, at the same time he finds himself bound by the modern technology of communications and transportation, which speed his world far beyond its original boundaries.
>
> Vine Deloria (1999a: 322–323)

Several years ago a high school teacher in Lawrence, Kansas, gave an assignment to her class to write about the topic "Where does your family come from?" A Comanche student in her class asked her how he was supposed to write about this topic because his people came from the high plains of North America. The teacher's response was, "Why don't you write about your ancestors crossing the Bering Strait?" Although probably well intentioned, this statement revealed surprising ignorance of and insensitivity to the true meaning of *Indigenous* and to the belief systems of Indigenous peoples.

The idea that human beings entered North America across the Bering Strait has become an unnecessarily contentious issue between scientists and Native American scholars over the last 10 to 15 years, not because Native Americans have trouble with the concept itself, but rather because this idea has been used by non-Natives to argue that "everyone is an immigrant to North America; Native Americans just arrived earlier." It should be no surprise to learn that similar arguments have been used against the Indigenous people of Australia, who may have the oldest human cultures that still exist and have certainly been in their land for at least 45,000 years and perhaps as long as 60,000 years (Mulvaney and Kamminga 1999). European invaders in Australia originally tried to claim that Indigenous Australians had arrived no more than 400 years before Europeans, and then modified this to a few thousand when evidence contradicted their earlier idea. A good review of this issue can be found in Chapter 13 of Carroll (2009).

One Indigenous scholar who spent many years debating this point was the Hunkpapa Lakota Vine Deloria Jr., son and grandson of preachers, and one of the most influential Native American scholars of our time. Deloria's writings on Indigenous perceptions of politics, history, religion, and education have been widely read and discussed. Deloria was without doubt an important scholar on Indigenous issues and philosophy, and is perhaps best known for his observation and discussion of a fundamental difference in the way that Western and Indigenous philosophical traditions view history and the role of humans within history (Deloria 1992, 1999a). Deloria's basic argument was that the Western tradition is temporally oriented—that is, history is viewed as a sequential series of events that show consistent progress, yet modern society relies upon accounts of events that happened thousands of years ago for moral and social instruction about how best to function in the world of today (Deloria 1992, 1999b).

In contrast, Deloria argued that Indigenous peoples view the world spatially, drawing their sense of identity from the local places where their cultures have evolved (Deloria 1992, 1999a, 1999b; see also Basso 1996; Pierotti and Wildcat 2000). This relates clearly to the ideas about how Indigenous traditions involve local communities that are based on relationships between humans and nonhumans, as I have presented in previous chapters.

According to Deloria (1992), temporally based thinking assumes that time proceeds in a linear fashion, and that in order to continue to be "right" we must constantly strive to be "better" than the previous point in time from which we came. In order to learn how to be better, it is important to study the "mistakes" of the past, that is, human history is "progressive." Therefore, when temporally oriented individuals encounter cultures that they consider to be "behind the times," such as Indigenous cultures, they strive to "advance" them. When the "primitives" refuse this help, this means that they fail to make "progress." As stated by another Lakota scholar, "[the] solution to this lack of ability to coexist with Indians was to locate Indians as far as possible from Euro-American society, out of the way of 'progress'" (J. Marshall 1995).

From Deloria's perspective time is not absent as a concept, but it is much less emphasized in cultures that employ spatial thinking. It is not that Indian languages have no word for time, as is sometimes claimed, but instead that time does not dominate the way the culture understands itself and the moral rules and authorities that determine codes of conduct, and the present is not assumed automatically to be superior to the past. Time clearly exists in the sense that events are perceived as having taken place in the past, but the exact date or time when such events occurred is of little consequence. Instead, what is important is understanding the meaning of these events to your cultural tradition, and that this tradition can best be exemplified by experiencing the physical location and ecological surroundings where the actual event occurred (Basso 1996). As a result, success is not defined by becoming better than those before you, but by following tried and true ways that have been shown to work repeatedly. In such traditions, survival and persistence provide the verification of the knowledge (Alessa 2009).

The primary method of historical instruction in a spatially oriented culture is to take younger members of the group to the places in which specific events occurred, and to tell them stories of these events so they can be visualized in a spatial sense (Basso 1996). Since the time at which such events occurred is not important, what does matter is that the people to whom these things happened were men and women just like the people of today, no better, no worse, hence there is no assumption of historical progression. Learning traditions and history in such cultures is facilitated by stories, which incorporate features of the landscape and local ecology. Places are given names that are considered to be important and of high value. In his chapter "Water Lies with Mud in an Open Container" Basso emphasizes the fact that when you say the name of the place, you need to say it right, the way that the ancestors did it. You need to recognize the lesson in the story and make sure you facilitate the transmission by calling it by the name it was given, because as the Apache elder explains to him, "It's matter" (Basso 1996: 10).

One important difference between spatial and temporal thinking relates to the manner by which information is communicated. The Western "literate" cultural tradition thinks that material that is written takes precedence over any other means, which renders oral performance as secondary and marginal (Le Guin 2004: 200). What they fail to realize is that oral performance uses time and space in a manner that allows it to create its own special relationship between speaker and listener, "a community of body and mind . . . that formal oratorical event can be echoed, can be shadowed, can be recollected, by films and recordings. Images of it can be reproduced. But it cannot. An event does not happen more than once. We do not step twice into the same river" (Le Guin 2004: 200).

Despite this difference in mode of communication, in reality the dichotomy between temporal and spatial thinking is not discrete, but more of a continuum. As Indigenous "spatially based" philosophies incorporate time, so do Western "temporally based" traditions incorporate a sense of place. This explains the historical sites and landmarks that are recognized throughout much of America, and more significantly, the wide range of historically and spiritually significant locations that are widespread throughout Europe and Asia. Ironically, Euro-Americans do not create or use sacred sites in America, possibly because they feel alienated from the "place," although the Vietnam Memorial in Washington and Ground Zero in New York may now be sacred locations, as they touch the spirit of the American people.

Ultimately the difference between these perspectives lies in the way in which they approach "truth" and the understanding of reality: the metaphysics and epistemology of the culture. Western philosophical traditions that Deloria links to "temporal" thinking tend to consider truth as universal or global and independent of the location at which it is defined. One goal of the Western scientific enterprise is to produce truths that are reproducible, again and again, and these truths are assumed to be independent of the location in which they are observed (see also Alessa 2009).

In contrast, "spatial" thinking sees truth as contingent—that is; it can be applied either to new physical locations, or to locations that have undergone significant physical change. At its essence truth is local rather than global, it acknowledges something that has worked previously, but what is assumed to be true may also change from day to day as the environment directs. As Le Guin describes it, "An event does not happen twice, yet regular recurrence is the essence of cyclic time. This year's spring is not last year's spring, yet spring returns" (Le Guin 2004: 201). Thus, a major benefit of spatial thinking is its adaptability to new or changed environments; it is possible to re-create stories as time progresses and thus to redefine "truths." The story is not forever fixed on the page, but can be altered in response to the new conditions that this new spring may present. As a consequence the temporal way of thinking views truth as absolute and unchanging, whereas spatial thinking views truth as pragmatic and malleable.

It may be more useful to discuss these concepts, not simply within the context of temporal and spatial, but, as I argued in Chapter 1, by using the terms *global* and *local*. Without doubt both time and space are accounted for in each philosophical approach. Within local concepts, place is emphasized, and the lessons received are in reference to the recent past. In a pragmatic sense this means, "not too long ago, this truth worked, therefore it probably works now." Within global concepts, time is emphasized; knowledge (truth) is cumulative, which is what allows cultural traditions to "progress." The lessons received are considered to be generalizeable and independent of people and places. In an absolute sense this means, "At a certain point in time, this truth was discovered, and thus it is now recognized as an undeniable law."

These statements represent the extremes of this continuum. To bridge the two, the scale at which the implications are to be understood must change. Local concepts work at fine scales, which are the scale at which individual interactions and relationships occur (Barsh 2000; Alessa 2009). Thus they are strongly linked to ecology and the idea that "all things are connected." Scaling upward or downward involves seeking truths that will work for not only you, but for you, me, and hopefully everyone else. In such cases it is necessary to take into account your special situation as well as those of other individuals. Further scaling up then proceeds from small to large groups, to countries, continents, the world, and presumably to both the atom and the universe, where Western physical scientists have found a great many laws that they regard as undeniable and irrefutable.

At each level more and more information must be considered. At the same time information and principles that are shared between groups or individuals becomes less and less and converges on the supposedly "undeniable laws." Global concepts work best at the largest and smallest scales, such as that of the universe or of molecules (Lewontin 2001; Alessa 2009). An example of this type of thinking can be illustrated by attempts to relate fundamental laws from the physical sciences, such as laws of thermodynamics, to biological phenomena, such as evolution and speciation (e.g., Brooks and Wiley 1986), which seems to be a response to creationist arguments that evolution violates the second law of thermodynamics (Petto and Godfrey 2007).

Where you position yourself on this continuum can translate into very different perceptions of social structure. In local concepts (spatial thinking) the perception of pragmatic truth proceeds to social structures and cultural traditions that emphasize necessity and intimacy, including a close relationship to the natural world (all things are connected). This is linked to the idea that Indigenous peoples understand the world based on relationships (Barsh 2000). In such cases, whether your perception is "subjective" or "objective" is not important. In contrast, when employing global concepts (temporal thinking), the orientation toward perceiving absolute truth emphasizes improvement and expansion but also requires emotional separation from what you observe, so that ideally you can attain an "objective" distance.

This supposedly "objective" dimension sets up in opposition to the idea that TEK lumps together empirically based observations and conclusions with spiritual/religious/philosophical ideas and assumes that these two kinds of knowledge are inextricably linked. Western science assumes an objective distance between observer and observed, whereas Indigenous observers do not concern themselves with objectivity. The problem is that this objectivity may not be as rigorous as is claimed. I have been a scientist for some 30 years and have observed a lot of subjective or at least non-objective behavior on the part of my colleagues. As one example, ecology is rife with economic metaphors that many scientists treat as literal truth, for example, producer/consumer, competition, optimal choice theory, etc. (Worster 1992, 1994). I have observed discussions about ecology and evolution between American and Russian scientists that basically mirrored the economic systems of their nations back in the 1970's. In Chapter 4 I mentioned the bias towards characterizing animal social systems only on the behavior of males that lasted until the early 1980's. Having been a scientist for most of my life I know that scientists are just like other humans, and bring biases, values, and even superstition into the way they live their lives, and this bleeds over into how they conduct their research. I am in no way saying that they are dishonest, but simply that every person's personal beliefs and values inform the way they conduct all aspects of their lives. As I pointed out in Chapter 6, many contemporary scientists are admitting the complexity of behavior among nonhumans that allows a more "subjective" view (Goodall 1986, 1990, Goodall and Bekoff 2003; Bekoff 2000, 2002, 2007; de Waal 1995, 2001, 2009).

SCIENCE AND VINE DELORIA, JR.

These new approaches mentioned above are more consistent with ideas developed by Deloria and could be used to provide a basis for reconciling modern evolutionary theory with TEK. Unfortunately Deloria, along with some other Indigenous scholars who write about Indigenous spirituality and religion (see chapter 5), failed to consider the broader scientific and philosophical implications of Indigenous belief systems, such as the idea that if the local environment changes, both the local human culture and members of the local ecological community are likely to change in response to the changed conditions, thus showing evolutionary change in a scientific sense (Pierotti and Wildcat 1997a, 2000; Mesoudi et al. 2004).

Despite the potentially powerful connections between Indigenous thought (including Deloria's own ideas) and modern evolutionary theory, as I pointed out in Chapter 5, there is a tendency of Indigenous scholars to argue that they do not believe in evolutionary change (Deloria 2002; Atleo 2004; Tinker 2004). These individuals argue that evolution is not relevant to modern Native Americans. I believe this is a case of throwing the baby out with the bathwater, which has led to inconsistencies in recent Native American scholarship. In order to understand the problems inherent in this line of reasoning, it is important to examine the underlying premises in some detail.

Deloria recounts the experience of defending Lakota land claims in a courthouse in South Dakota when a woman of European ancestry commented to him concerning the Bering Strait theory, "Well, dearie, we are all immigrants from somewhere." Deloria's (unspoken) response was, "It makes [a] difference whether we came 100,000 years ago or just out of the boat a couple of generations back" (Deloria 1995: 83--4). Deloria's point is well taken, even if one might choose to quibble with his choice of time frame. My personal feeling is that the best response to such self-serving Euro-American discourse is to quote another eminent Lakota, the Sicangu author Joseph Marshall, who discusses the Bering Strait controversy as follows: "The original stories among the many peoples of Turtle Island [North America] do not bother with when. Instead, many such stories deal with the obvious fact that we are here and that we have always been here. When a moment or an event happened so long ago that it has ceased to exist in collective memory, it then begins to exist—as my grandfather used to say—on the other side of memory. In such an instance, always becomes a relative factor, and what emerges as far more important is first" (J. Marshall 1995: 207).

In any case, it was appropriate for Deloria to be offended by this nonsense of dismissing land claims by citing the Bering Strait hypothesis as an example of immigration. Deloria established himself as an important thinker on all things Indian in the 1970s and 1980s. However, in the 1990s Deloria decided to re-imagine himself as a critic of Western science, especially with regard to the issue of whether or not the Bering Strait hypothesis for the peopling of the Americas was supported by solid evidence. Deloria addressed this topic in detail in his book *Red Earth, White Lies* (1995), the title of which illustrates his position. Deloria next took on the issue of evolution versus creationist thought in the last book he published during his lifetime: *Evolution, Creationism, and Other Modern Myths* (2002).

I find Deloria's arguments in both of these books troubling. He uses numerous statements indicative of a "young-earth creationist" of European ancestry, that is, that Mother Earth is so new that she came into existence around the time that humans, including Native American peoples, arose (Deloria 2002). The young earth creationist point of view is that the universe was created in six 24-hour days, and that the earth's geology and life forms were shaped by a literal, historical flood, which is why it is sometimes referred to as "Flood geology" (Shermer 2006; Scott 2007).

From my point of view, the perspective taken by Deloria amounts to an insult to our planet, which has been in existence so long that humans are a very new species. Oddly, Deloria himself has acknowledged in his other writings that human beings were "the last creatures to be created and the youngest of the living families" (Deloria 1999b), yet like creationists of European ancestry, in 2002 he did not seem capable of imagining a world without humans.

Deloria's basic contention in his last book is that evolution equals Western science and functions as a new religion (Deloria 2002). The only apparent difference between what Deloria writes and the writings of fundamentalist Christians is that he does not use the term *secular humanism* to describe this new "religion." One of the strangest statements made in *Evolution, Creationism, and Other Modern Myths* is that now that Deloria sees science as having replaced religion, this means that "Nature" has replaced "God." This is a strange statement for an Indigenous scholar to make, especially one who has written extensively on the difference between Western and Indigenous perspectives on religion. As I have discussed in previous chapters, in most Indigenous religious traditions Nature is treated as if it were equivalent to the creator (or creation). Deloria himself has stated:

> No element in of life [*sic*] can go unattached from human society. Thus lands are given special status because they form a motherhood relationship with the peoples who live on them . . . The true meaning of *the motherhood of the land is that, like a mother, it shapes and teaches our species*, and according to the peculiarity of the area, produces certain basic forms of personality and social identity which could not be

> produced in any other way . . . With respect to other life forms this attitude manifests itself in what one could call 'kinship' cycles of responsibility that exist between our species and the other species . . . This transformation brings out a dimension of life common to Indians, but unique and unsuspected by non-Indians. (Deloria 1999a: 131, emphasis added)

If this is what Deloria believes, he is perilously close to equating Nature (or land) with the creative force. As I pointed out in Chapter 5, according to nineteenth century Dakota, "our mother, the Earth . . . conceived and brought forth life, both vegetable and animal" (Eastman 1980: 123). To be intellectually and philosophically consistent and to follow Lakota tradition, Deloria should embrace the concept that the creator (God) is equivalent to Nature in many important ways.

A major flaw with Deloria's critiques of science is that he contented himself with reading and critiquing only popular accounts of evolutionary biology. By his own admission, Deloria states, "I became an avid reader of *popular* scientific books, wanting to know as much as I could about the world in which I lived. Gradually I began to see a pattern of nonsense in much scientific writing" (Deloria 1995: 9; emphasis added). Having reached a conclusion based on popular accounts, Deloria chose not to investigate original sources to see if the problems he identified lay with the popular writers or with the science itself. Instead Deloria chose to rely on popular accounts, as well as articles from newspapers and magazines, which allowed him to pick and choose what he decided was a useful argument. This choice of venue probably comes from Deloria's formal training in law and theology, which means that he argues as an advocate and looks for any material, however questionable, that can be used to attack an opposing point of view. This approach, however, does not serve him well as a serious critic of science.

In the courtroom, popular accounts are given credence because, in the case of newspaper articles in particular, they may represent all of the written material concerning an event. In science, however, there is primary literature—peer-reviewed journal articles and scholarly texts written by individuals who actually conducted the research or developed the ideas that they report. Primary literature is the best source because it represents the direct results of scientific study and often includes extensive tables or figures representing the actual data used to support the arguments. Access to the data used to draw conclusions allows the reader to independently judge the quality of the work.

The next level is secondary literature—articles in *Scientific American, Discover*, or *Natural History* or books written about a topic by scientists or about science by nonscientists, such as the popular writings of Steven J. Gould, E. O. Wilson, or Carl Sagan, books by science writers such as Jonathon Weiner or Carl Zimmer, including in the social sciences and humanities even works such as Shepard Krech's *The Ecological Indian* (see Chapter 8). As an obvious example, the book you hold in your hands is basically a work of secondary literature with some original ideas included. Therefore, secondary literature can be useful, but it needs to be fact-checked against primary literature to ensure that the scientific research is accurately represented.

Finally there is tertiary literature, which is represented primarily by basic textbooks or popular accounts written by individuals who are not experts in the field, such as most of the reviews of *The Ecological Indian* discussed in Chapter 8. In order to judge its quality, tertiary literature needs to have its statements checked for veracity because information is often simplified or misrepresented.

In his critiques of science Deloria appears to rely entirely on secondary and tertiary literature. It is rare to find a case in his books and articles on science where he cites an article from the primary literature. Nonetheless, Deloria's critique of science is not completely off the mark. As I indicated above, science is a human endeavor, conducted by human beings who are egotistical, biased, emotional, and rarely as objective as they would like the world to perceive them as being. There has been bias in science at least since Aristotle postulated that men had more teeth than women and then refused to count the teeth of his wife to check the accuracy of his statement. If Deloria had gone to the primary literature and reviewed it carefully and thoughtfully, he might have produced a truly insightful critique of the cultural biases inherent in Western science and suggested new directions for scientific endeavor to explore using techniques and ideas developed by Indigenous peoples.

SCIENCE, FAITH, AND PLEISTOCENE OVERKILL

As one obvious example, the evidence for the "powerful and entrenched" Bering Strait idea (E. Marshall 2001a) is much more problematic than is generally acknowledged, but Deloria missed this literature completely. The Bering Strait concept is based on surmise as much as on actual evidence and is highly vulnerable to a solid critical analysis (Grayson 1991, 2006; Grayson and Meltzer 2002, 2003). The basic idea contends that humans moved into the Americas across the Bering Strait during the last glaciation and then moved down into the North American continent proper through an ice-free corridor east of the Canadian Rockies (O'Neill 2004). The Bering Strait hypothesis may be more accurately described as the "Clovis Hypothesis" because it is closely tied to, and dependent on, the idea that the earliest human site in the Americas is indicated by the appearance of distinctive fluted spear points first discovered around a site near Clovis, New Mexico, dated to 11,800 years bp (before the present). Such spear points were subsequently found in numerous sites throughout North America (E. Marshall 2001b).

Another, even more controversial idea linked to the timing of the "Clovis-Bering" idea is "Pleistocene Overkill," or the idea that human hunters drove much of the American megafauna into extinction during the period 10,000 to 12,000 years bp (P. Martin 2005). This idea has been extended into what is now called "Aboriginal Overkill" (Alvard 2002; Kay 2002a,2002b), which argues that even relatively recent Indigenous Americans have had devastating impacts on wildlife, although not as devastating as those imposed within the last few centuries by the European invaders (Broughton 2002; Mann 2005). The linked ideas of Bering-Clovis and Pleistocene Overkill (henceforth referred to as Bering-Clovis/Overkill, or BC/O) have become part of the established dogma of archaeology (E. Marshall 2001a). In fact, Paul Martin, originator of the Pleistocene Overkill hypothesis, found it necessary to include a self-serving commentary that both defends the combined ideas of Clovis, the Bering Strait, and Pleistocene Overkill and criticizes Deloria's comments on this linkage (P. Martin 2005: 145–147).

The reason that the overkill idea appears to be significant, and leads to pairing of these concepts from an Indigenous perspective, is that the idea that North America was peopled across the Bering Strait has become the basis of a completely spurious argument made by many people of European ancestry, stating that the Indigenous peoples of the Americas are "immigrants just like Europeans, only earlier" (see above). Combined with this argument in recent years have been arguments advanced by political conservatives, and given support by Krech (1999a; Chapter 8), that Indigenous Americans were "not good ecologists" because in the past they are assumed to have driven numerous species to extinction. "Paul Martin . . . proclaimed that 'man and man alone was responsible' for the extinctions. . . . Branding the ancient Indians—the so-called Paleo-Indians—as superpredators. Martin likened their assault on Pleistocene animals as a blitzkrieg; evoking the aggressive, assaulting imagery of the Nazi war machine" (Krech 1999a: 29).

Three major problems exist with the linking of these concepts and, in fact, even with the BC/O concepts themselves. First, numerous sites scattered widely throughout the Americas are indicative of human occupation much older than at the Clovis site (E. Marshall 2001a; Goebel et al. 2008). Second, supporters of the BC/O idea tend to ignore the possibility of humans moving into North America by any means other than the ice-free corridor. This is important in terms of timing, because this corridor did not open until approximately 13,000 years bp (E. Marshall 2001b; see also Guthrie 1990; Lemonick and Dorfman 2006). The final, and perhaps most serious, problem with the BC/O ideas is that the Pleistocene Overkill hypothesis assumes that prey species, including mammoths and mastodons, were naïve, inexperienced, and highly vulnerable to humans as a new predator in their ecosystem.

The probable existence of multiple sites older than Clovis seems to have become a political rather than a scientific argument among advocates of the BC/O idea. The arguments presented by BC/O advocates serve as good examples of the subjectivity of some issues in science. At least six sites in North America, as well as several in South America, have been dated as significantly older than the Clovis site in New Mexico (E. Marshall 2001a). Every time a new site is identified, its validity is questioned, usually by C. Vance Haynes, emeritus professor of geosciences at the University of Arizona, member of the National Academy of Sciences, and not coincidentally, the individual responsible for dating the Clovis site in the first place. Haynes appears to have appointed himself defender of the Clovis site as the first major human site in North America, even though it is several thousand miles from the Bering Strait. By extension Haynes also serves as one of the primary defenders of the Bering Strait argument. Paul Martin, who developed the overkill hypothesis and serves as its chief advocate, is also an emeritus professor of geosciences at the University of Arizona. This means that the two individuals largely responsible from pushing the BC/O arguments are long-term colleagues who have steadfastly resisted attempts to establish evidence suggesting that humans have been in the Americas much longer than their findings suggest.

This is a crucial point, because if humans co-existed with the megafauna for thousands of years prior to their supposed extinction around 10,000 years bp, this would strongly suggest that hunting by early American humans was not the primary factor responsible for these extinctions (Broughton 2002; Eswaran 2002; Pierotti et al. n.d.). If the megafauna co-existed with humans for a considerable period of time, they were almost certainly not naïve about humans as predators when they went extinct. It is possible that the most significant relationship involved in BC/O ideas does not occur in the data, but instead in the fact that Haynes and Martin work together and have a powerful vested interest in censoring ideas that would render much of their scientific reputations insignificant. The most recent review of the timing of human arrival in the Americas was published in the prestigious journal *Science* (Goebel et al. 2008) and combines both genetic and archaeological evidence to conclude that humans occupied Beringia (the portion of Siberia and Alaska that are geologically linked) around 32,000 years bp and that there were no barriers to passage at this time to occupation of North America. There are at least four sites in what is today the United States that show evidence of probable human activities as far back as 22,000 years bp (Goebel et al. 2008: 1500). According to these scholars there is conclusive evidence that humans were in North America by 15,000 years bp at the latest and that these people almost certainly moved down the Pacific Coast using boats. This suggests strongly that although humans may have later used the interior corridor along the Rocky Mountain range, humans were in the Americas several thousand years earlier than is argued by the BC/O hypothesis. This also means that the megafauna would have had thousands of years of experience with humans prior to their extinction.

The most widely accepted site older than Clovis is Monte Verde in south central Chile, dated to at least 12,500 years bp (E. Marshall 2001a). What is especially significant about this site is its location, several thousand miles south of the Clovis site (not to mention the Bering Strait). This date and location suggest clearly that to have a well-established culture in southern Chile by 12,500 years bp, humans had to have been in the Americas well before that time. Other sites have been found in North America that have been dated from 11,000 to 22,000 years bp. In each case Haynes has gone on the attack, questioning everything from stratigraphy to collection techniques to the methods of dating employed. It seems possible that Haynes is more interested in maintaining the priority of his own work than he is in increasing knowledge. This is an illustration of what I mentioned earlier about science being a human endeavor, egotistical, potentially biased, and rarely as objective as we are led to believe.

One aspect of the BC/O controversy that seems to go largely unmentioned by its advocates is that during the ice ages, sea levels were much lower than they are today, for the simple reason that much of the world's water was tied up in glaciers that covered much of Eurasia and North America. This lowering of sea level is what allowed the Bering Land Bridge (which might more accurately have been described as the Bering Great Plains since the "bridge" covered several hundred miles north to south) between Asia and North America to exist in the first place. This point goes unrecognized, not only by the archaeological canon, but unfortunately by Indigenous scholars as well (Parfit 2000; Goebel et al. 2008; Carroll 2009).

For example, Deloria comments that he was unable to find any scholarly article proving that Paleoindians had ever actually crossed the Bering Land Bridge. This is unsurprising, considering that most of the land where this evidence might be found is today under the Bering Sea. A pointed out above, recent study has shown evidence of humans in Beringia for at least 32,000 years (Goebel et al. 2008). It may not be surprising that nonscientists fail to make this link. More troubling is that until recently it seems not to have occurred to archaeologists that much of what made up the Pacific Coast of both eastern Asia and western North America during periods of major glaciation are today under the Pacific Ocean. Only recently have studies been conducted on marine archaeology along the Pacific Coast of Canada. It is significant that these studies indicate that the presence of humans in the Queen Charlotte Islands (Haida Gwai of the Haida people) dates to 12,000 to 15,000 years bp (Parfit 2000). This timing is consistent with that of almost all of the pre-Clovis sites; however, it is inconsistent with the BC/O hypothesis. The implication of these results is that humans moving into the Americas used coastal routes rather than the interior ice-free corridor, and they may have co-existed with megafauna in North America for as much as 15,000 years prior to the extinction of these creatures (Guthrie 1990; Parfit 2000; E. Marshall 2001a, 2001b; Lemonick and Dorfman 2006; Goebel et al. 2008). This suggests that the fauna were used to human hunters, and that humans may have had little impact on megafaunal extinctions relative to major climatic and ecological changes.

The issue of extinction resulting from human overkill, except on islands, has always been problematic to anyone with a good understanding of ecology (Guthrie 1990; Grayson 1991; Elias et al. 1997). The period of maximum extinction is coincident with a period of maximal glaciation, which indicates strongly that climate change was almost certainly involved, as well as compression of habitats and the ecological ranges of many species as glaciers covered much of North America (Guthrie 1990), extending as far south as Kansas and Missouri. In addition, it is an ecological rule that except in models, predation is rarely involved in the extinction of prey species on continental landmasses, although it is frequently involved on islands (Shugart 2004), where prey species have adapted to conditions without predators, such as birds that evolved to a flightless state.

There is no obvious pattern to the species that went extinct in North America compared with those that survived and persist to this day (see Table 1 in P. Martin 2005 for a complete list of extinct species). If predation by humans was important, why did large dangerous carnivores, such as the sabre-toothed cat, American lion, American cheetah, short-faced bear, and dire wolf go extinct, while their close relatives the puma, jaguar, grizzly and black bears, and red and gray wolves persisted? Similarly why did some species of bison, mountain goats, peccary, and pronghorn go extinct, while closely related congeners (members of the same genus) survived? The ecological pattern that makes the most sense is that ecological specialists died out while more generalized relatives persisted, which suggests that ecological changes other than predation were most likely to be the causative factors (see also Guthrie 1990; Grayson 1991, 2006; Grayson and Meltzer 2002, 2003). The real enigma is the proboscideans: Why did mammoths and mastodons go extinct, whereas tropical elephants survived? Admittedly there are a few mastodon kill sites, of which Martin makes a great deal. He suggests that early American humans may have specialized on these large, social, intelligent, and potentially dangerous mammals. In my opinion, because of their intelligence, strength, and tendency to act as an organized group, elephants are much less likely to have been regular targets of hunters than ungulates such as bison, moose, wapiti (elk), deer, caribou and pronghorn, all of whom persist to this day.

As mentioned above, a more recent approach to examining possible impacts of hunting by Indigenous Americans has been characterized as "Aboriginal Overkill" (Kay and Simmons 2002). Scholars following this line of thinking expand upon the basic idea of "Pleistocene Overkill," extending it into historical times and arguing that the numbers of Indigenous Americans were so great that they must have had considerable impact.

This position is characterized as follows:

> Contrary to (the) prevailing paradigm . . . native people were originally more numerous than once thought . . . (they) were generally not

> conservationists . . . Instead native people took an active part in managing their environment. Moreover, changes wrought by native people were so pervasive that their anthropogenic, managed environment was thought to be the "natural" state of the American ecosystem." (Kay and Simmons 2002: xii)

Although I think there is some merit in this position and have made similar arguments in previous chapters, this approach also has problems. Advocates of Aboriginal Overkill rely for the most part on the results of theoretical models to support their arguments. Principal among these is "Optimal foraging theory," which is a body of models and concepts developed to explain prey choice in foraging nonhumans (Stephens and Krebs 1986). "Foraging theory," as the discipline is now known, has evolved well beyond the primitive models employed by the authors in Kay and Simmons (2002). These authors use only models for "Optimal diet choice" which are basically models of prey and foraging patch choice (Marginal Value Theorem) that assume that the sole goal of a foraging individual is to maximize the amount of calories they obtain from a single food item (Stephens and Krebs 1986). The basic logic underlying such thinking is that animal calories are more valuable (than plant calories) to native people because of their high fat content (Kay and Simmons 2002). They further argue that, based on studies of human hunters that "large mammals have the highest handling efficiencies," and that taking large mammals is much more efficient than hunting small mammals, birds, or fishing (xviii).

All this is fine, except that they then leap to the assumption that if hunting contemporary large mammals (using firearms) is the most efficient form of food acquisition, then hunting mammoths and mastodons (using spears) must have been a optimal form of foraging (Kay 2002a, 2002b). Foraging theory has moved well beyond simple diet choice models (e.g., Mangel and Clark 1988; Giraldeau and Caraco 2000). The models and concepts in these later books emphasized dynamic foraging decisions and "risk prone" versus "risk averse" foraging. The latter idea is based on the logical assumption that taking readily available or catchable prey is often better than seeking high reward but risky prey (Caraco 1981). The logic of assessing risk in foraging would suggest that hunting large, aggressive, highly social animals such as mammoths and mastodons, not to mention other extinct megafauna such as sabre-toothed tigers and short-faced bears, would be very risk-prone forms of foraging.

Most evidence supporting Pleistocene overkill consists of loose correlations and surmise about the timing of human arrival, which as described above, are dependent on presence and absence of ice, the alleged deadliness of Clovis spear points, and the supposed naïveté of prey species. The results of the one major study that is not simply guesswork and faith and examines how naïve prey respond to introduction of new predators was presented as a test of the Pleistocene Overkill hypothesis (Berger et al. 2001). Oddly, the results obtained seem to actually show the opposite of what was predicted. This study demonstrated unequivocally that the prey involved—moose (*Alces alces*) and American elk (*Cervus elaphus*)—appeared unfamiliar with wolves or bears at first, but once individuals of these species observed wolves or bears take a member of their species, they learned immediately to respond to them through avoidance and defense. They also taught their offspring to avoid these predators, which supports the argument I made against Krech's (1999a, 1999b) comments about early Indian hunting of bison and bison learning in Chapter 8.

The result of the Berger et al. (2001) study, which to me clearly does not support the Pleistocene Overkill hypothesis, amazingly is cited by Paul Martin (2005) as follows: "Berger's [*sic*] recent findings suggest that *at first contact* the American fauna would have lacked behavioral defenses against humans including the fear and alarm response necessary to inspire potential prey to fight or flee. . . . Very likely it would have taken longer for potential prey to learn to fear the new human predators than it did for the moose to learn to fear the wolves, *which their ancestors had known to be dangerous*" (Martin 2005, 139; emphasis added). This is the response of a person who studies fossils rather than living creatures—the behavioral assumptions implicit within it bear a closer resemblance to science fiction than to science. It also shows that not only the public, but even scientists are often prepared to uncritically believe what a scientist tells them, rather than to critically evaluate the evidence themselves.

It is interesting to note that Martin considers moose more likely to learn quickly than mammoths or mastodons, who like most proboscideans are much more "intelligent," or at least behaviorally complex, than cervids such as moose and elk. Even more interesting is Martin's apparent invocation of a sort of ancestral "memory" in moose, while ignoring the fact that mammoths and mastodons also had ancestral experience of human hunters in Asia, where they had coexisted for as long as 40,000 years, as well as the well-established ability of proboscideans to pass knowledge across generations (Payne 2000; McComb et al. 2001; Goebel et al. 2008).

This special pleading, along with Martin's misuse of the results of a scientific study, is the sort of thing that an effective critic might be expected to take on in his critique of the scientific validity of the Bering-Clovis/Overkill hypotheses. As one example, non-Native scholars examining the evidence have described the BC/O hypotheses as follows:

> Martin's recent writings suggest to us that he is no longer trying to approach this issue within a scientific framework. As we have noted, he explicitly maintains that the North American overkill position does not require supporting evidence. He is unconcerned that archaeologists "wash their hands" of his ideas. He criticizes the search for pre-Clovis sites in the New World as "something less than serious science, akin to the ever popular search for 'Big Foot' or the 'Loch Ness Monster."

> As one of us has observed elsewhere, *Martin's position has become a faith-based policy statement rather than a scientific statement about the past*, an overkill credo rather than an overkill hypothesis." (Grayson and Meltzer 2003: 591, emphasis added)

EVENTS BEYOND MEMORY: RED AND WHITE DISSEMBLING

Unfortunately Deloria made none of these points. He does point out that a number of tribes have traditional stories describing their encounters with "ice that reached to the sky" or "lands of perpetual snow and ice," which suggest that these peoples were in America prior to the advent of glaciation and entered America during interglacial periods (Deloria 1995: 98–99), but Deloria reveals surprising ignorance in other places. Referring to events that took place over the last 20 million years, he argues, "I can't imagine thousands, perhaps millions of horses and camels struggling to get through the McKenzie Mountains, perhaps stampeding up Skagway Pass, crossing the land bridge, and then being confronted with . . . rugged mountain ranges which they had to traverse before they found a home in the steppes of Asia" (Deloria 1995: 100). I hope this statement is meant to be taken tongue in cheek, because it appears to assume that neither the climate nor the landscape of North America has changed over the last 20 million years, during which the Bering Land Bridge has existed at various times.

This illustrates a serious problem: Both Deloria and Martin are inventing their own scenarios and then arguing that they must be true. Martin knows how to phrase his arguments like a scientist; therefore his credibility appears greater, even though it really is not. Deloria weakens his position through his failure to understand the age of the earth and that the surface of the earth constantly changes. During the Miocene and Pliocene, when major faunal exchanges between Asia and North America took place (and Beringia is the only place this exchange could have taken place, because unlike early humans—camels, horses, bison, mammoths, and wolves do not use boats), the world climate was much milder, and there was little snow or ice to traverse. Deloria does not seem to realize that these organisms were widely distributed during extensive periods of mild climate.

There is nothing wrong with Deloria criticizing, and at times even lampooning, the evidence used to argue that Indigenous Americans are "immigrants" and caused "Pleistocene Overkill." To do so effectively, however, he should have critiqued the evidence as was done by Western scholars (Grayson 1991, 2006; Grayson and Meltzer 2002, 2003). Instead Deloria throws out the baby with the bathwater. He attacks all science and seems to rely on an odd combination of pseudoscience and "Native American Creationism" to support his arguments. Thus Deloria not only fails to present an effective rebuttal to the views expressed by the Arizona old-boy network, but he even provides individuals such as Paul Martin the material to effectively refute him and make him seem naïve (Martin 2005: 146).

A major problem with Deloria's analysis of the problems with BC/O arguments was that he insisted on the literal truth of the metaphor "We have always been here" as a way of establishing the primacy and Indigenous validity of Native American peoples. Given this metaphor and Deloria's impassioned defense of it, the only reality that made sense to his arguments was that the earth was much younger than scientific evidence suggests. It is telling that fundamentalist Christians also use the idea of a "young earth" to support their belief system as well as to oppose evolutionary theory, both Darwinian and Neo-Darwinian (Shermer 2006; Petto and Godfrey 2007).

Late in his career Deloria attacked evolutionary thought as if he considered it to be potentially harmful to TEK and Indigenous sovereignty (Deloria 2002). This parallel is more than mere coincidence. I personally attended a "Traditional Knowledge Conference" to which Deloria invited Donald Patten, an avowed Young Earth creationist of Euro-American ancestry, to be a featured speaker, and he cites Patten as an authority in both of his books on science (1995: 169–173; 2002: 107). An important point to realize is that Patten is very much a "Flood geologist" style Young Earth creationist (Petto and Godfrey 2007). At that conference, I asked Patten about the timing of events in North America, and he told me that the Grand Canyon was created instantly by floods and that giant crocodiles existed in the American Southwest that could have exterminated the megafauna. When I pointed out that the crocodilians to which he referred had existed during the Mesozoic period, over 100 million years ago, Patten's reply was "No, the earth was created in the last 10,000 years." Clearly Deloria did not believe this to be the case; otherwise there is no sense to his statement I cited above: "It makes [a] difference whether we came 100,000 years ago or just out of the boat a couple of generations back" (Deloria 1995: 83–84). Still, Deloria presented Patten as a viable critic of Western science, and encouraged other participants in the meeting to take Patten seriously.

As a consequence, Deloria's critiques of European science fly in the face of much of Indigenous knowledge—the ideas that all life forms are related and that most nonhuman species existed prior to or as "elders" to humans (see Chapters 2, 4, 5; Pierotti and Wildcat 1997a, 2000). Ironically, one of the strongest pieces of evidence supporting the antiquity of human presence in the Americas is the evidence from DNA, which suggests that humans have probably been in the Americas for at least 15,000 years (C. Mann 2005; Goebel et al. 2008). Deloria ignores such evidence, which, unlike the Young Earth creationist arguments of Donald Patten, actually supports his point of view. He became so anti-science that he opposed evolution, which is based in large part on understanding patterns in DNA.

As I argued in Chapter 5, Deloria seems to accept the idea of "relatedness" among all life forms but struggles with the concept of "creation." By doing this, Deloria actually seems to reject and ignore Indigenous tradition and possible ways of understanding in his effort to defend one narrow aspect of Indigenous tradition that almost certainly results from colonized thought imposed by Christian missionaries.

As a way to possibly understand this decision, it is worth considering Deloria family history (Treat 1999; Deloria 2000). Vine's great grandfather, François des Lauriers, was Dakota and a healer on the Yankton Reservation, where he was known as Saswe (a Dakotization of François). Saswe was baptized in 1871 after losing two of his three wives. His first son, Tipi Sapa, assisted Saswe in his work as a healer until he decided to become an Episcopal priest at the age of sixteen, at which time he was baptized Philip Joseph Deloria. After his father's death, Philip Deloria helped found *Wojo Okolakciye* (The Planting Society), an organization promoting Christian fellowship on the Standing Rock Reservation, where he was appointed supervisor of all missionary work in 1892. Philip's only son Ohiya was christened Vine Victor Deloria in 1901. In 1931 Vine Deloria Sr. graduated from an Episcopal seminary and was assigned as a pastor to All Saints Mission on the Pine Ridge Reservation where his son, Vine Victor Deloria Jr., was born two years later in 1933. Vine Deloria Jr. attended a Lutheran seminary in Rock Island, Illinois. Intending to follow his grandfather and father into the ministry, he received a graduate degree in theology in 1963 (Treat 1999: 5–8).

Given this background it is not surprising that Deloria was strongly influenced by Christian thought and ideas. Much more ironic and troubling is that Vine Deloria Jr., the architect of the concept that Western and Indigenous cultures differ from one another, the former being "temporally oriented," and the latter "spatially oriented" (see above and Deloria 1992, 1999a), should in his defense of Indigenous traditions rely on one of the most temporally based Western concepts—the idea that the world did not come into existence until human beings were created. In a recent meeting of Indigenous scholars in Oklahoma, Vine's son, Philip Deloria, made the statement that "Vine defended Indigenous spirituality without being actively engaged in it."

As a consequence, Vine Deloria appears confused and inconsistent on this point. He has argued that "the primary focus of many tribes placed human beings as among the last creatures who were created . . . our job was to learn from other older beings, and so pattern ourselves after their behavior" (Deloria 1999b). In a work published the same year, he states that "American Indians look backwards in time to the creation of the world and view reality from the perspective of the one species that has the capability to reflect on the meaning of things" (Deloria 1999a: 130). The inconsistency between these two statements in Deloria's thinking suggests that he had difficulty reconciling his views.

The resolution to this apparent conundrum is, as has been argued in previous chapters, that creation stories of the various Indigenous peoples of North America refer not to the origin of humans but to the origins of specific cultural traditions (see also Mesoudi et al. 2004). Understanding the way in which origin stories were used within Indigenous cultures, and using Deloria's construct of spatial versus temporal worldviews, provide a way out of the dilemma that trapped him toward the end of his career. If he had chosen to critique the BC/O hypothesis through an examination of the scientific evidence and had then rephrased the argument within an Indigenous context, he would have been able to provide a powerful defense of Indigenous cultural traditions. Unfortunately his Christian religious background caught him in the trap of disavowing all evolutionary thought, regardless of whether or not it was in conflict with Indigenous traditions. It is unfortunate that a deeply thoughtful and intelligent individual, who once wrote that "I have in my lifetime concluded that Christianity is the chief evil ever to have been loosed on the planet" (Deloria 1999b: 146), should end up allying himself in 2002 with the point of view he had thoroughly repudiated just three years previously.

10 Who Speaks for the Buffalo?
Finding the Indigenous in Academia

> We went to Geneva as representatives of the indigenous peoples of the Western hemisphere, and what was the message that we gave? "There is a hue and cry for human rights," they said "for all people," and the indigenous people said "What of the rights of the natural world? Where is the seat for the Buffalo or the Eagle? Who is representing them here in this forum?"
>
> Oren Lyons (Onondaga; 1984)

To be indigenous means that the entity in question occurs naturally in a specific area or environment. This concept carries the implication that the organism or culture under discussion evolved within the environment that is under study. With Indigenous cultures this means that the set of conditions found within that environment shaped the traditions and knowledge set of the culture (Basso 1996; Pierotti and Wildcat 1997a, 1997b, 1999b, 2000). People who are indigenous to a specific locality look at the environment around them to get a sense of their place in history, and depend upon the animals and plants of their local environments for companionship, as well as for food, clothing, and shelter.

Given these links to the natural, nonhuman world, there exists a potential problem in the intellectual and academic disciplines that are involved in the study of the Indigenous peoples of North America. Such disciplines seem to eschew the sense of place that is inherent in an Indigenous identity (Deloria 1992; Cajete 1994; Basso 1996). In contemporary academia the study of the Indigenous peoples of the Americas includes practically all disciplines but the natural and physical sciences. The absence of the natural sciences from Indigenous or Native American studies programs is especially poignant when we remember that Indian people often defined identities and traditions in terms of their connections to and relationships with with the natural world (Lyons 1984; Deloria 1992; Cajete 1994; Anderson 1996; Pierotti and Wildcat 2000; Alessa 2009).

In previous chapters, I have shown that Indigenous scholars from other disciplines can commit serious errors when they try to discuss or interpret science. In addition, such scholars also have difficulty in refuting arguments from the Western scientific establishment. As an example, Vine Deloria Jr. struggled when he attempted to criticize issues that supposedly arise from science. The arguments that Deloria opposed are highly vulnerable to criticism, but he had no idea of how to rebut them effectively; instead he resorted to ridicule and hyperbole, which unfortunately cost him credibility within academia.

Few Indigenous scholars have any substantial training in science. Out of a list of 378 Native Americans known to have received PhDs by 2004, only 24 (6%) had degrees in Biology or related fields (including Botany, Medicine, Biological Psychology, and Vertebrate Paleontology). Three of those 24 individuals are no longer alive, including Charles Eastman (Ohiyesa) the Dakota physician and scholar who wrote eloquently about his people and their cultural traditions (Eastman 1971, 1980, 1991). Only 12 of these individuals have obvious associations with universities, and two others are employed by the National Institutes of Health (source: list of Native Doctorates compiled by Cecil Thomas, 8 June 2004). Given the theme of this book, it is of interest to note that only six of these 24 Indigenous Biology PhDs are in ecology or evolutionary biology. This means that there are very few mentors or role models for Native American students in Biology, Ecology, or Environmental Science.

The most likely explanation for the overall low numbers of scientifically trained Indigenous PhDs lies in the realization that the concepts and set of observations described as Western science is often considered to constitute a hostile environment for most Indigenous students (Deloria 1995, 2002). Despite this situation, however, there appears to be a growing consensus, which is the theme of this book, that Indigenous knowledge about the natural world and its processes is extremely accurate. More importantly, TEK or Indigenous knowledge describes and recognizes real processes and phenomena that mainstream science is only beginning to acknowledge. As these processes are rediscovered by Western science they are changing the nature of science and knowledge (Martin 1999; Pierotti and Wildcat 1997a, 1997b, 1999b, 2000; Barsh 2000).

The power of this knowledge base of Indigenous peoples comes from its emphasis on local knowledge, because as I discuss in earlier chapters, it derives from the web of relationships between humans, nonhumans, and landforms within a particular locality (Cajete 1994; Basso 1996; Silko 1996; Pierotti and Wildcat 1997a, 1997b; Barsh 2000). "In the traditional [way of knowing], there is no such thing as isolation from the rest of creation" (Deloria 1990). Indigenous worldviews are described as "spatially oriented," in contrast to the "temporally orientated" worldview representative of Western culture (Deloria 1992). Within a spatially oriented worldview, ecosystems are conceived of as a web of social relationships between the family, clan, or tribe and the nonhuman elements with which they share a place (Silko 1996; Barsh 2000; see also Chapter 2).

A common general philosophy and concept of community appears to be shared by all of the Indigenous peoples of North America, which includes: 1) respect for nonhuman entities as individuals, 2) the existence of bonds between humans and nonhumans, including incorporation of nonhumans into ethical codes of behavior, and 3) the recognition of humans as part of the ecological system, rather than as separate from and defining the existence of the system. The ideas that "All things are connected" and "All things are related" emerge from these philosophical concepts. As I have argued in previous chapters, these ideas serve as both the intellectual and the spiritual foundation for an Indigenous theory and practice of politics and ethics centered on natural places and connection to the natural world, which is based upon empirical observations resulting from patient observation of the natural world and its patterns.

It would seem logical therefore that academic departments, programs, or centers dedicated to "Indigenous," "Native American," or even "ethnic" studies would incorporate academic disciplines that deal with the sciences of geology (earth science) and biology. This situation does not prevail for two reasons. First, as mentioned above, there are few Indigenous academics with advanced degrees in these disciplines. There is a need for an intellectual insurgency movement in the sciences, and what better place is there for such a movement to emerge from than American Indian, Native American, and Indigenous studies programs. Traditional Indigenous worldviews and knowledge systems shun the naïve dichotomies of Western thought, and one of the most pernicious of those concepts is that a dichotomy exists between sciences and the humanities.

The second reason for the absence of scientists from such programs is that, despite TEK, Indigenous traditions, and efforts that are supposedly dedicated to decolonizing Indigenous academics, the study of Indigenous people and their cultures appears to be located almost entirely within disciplines that separate humans from the natural world. It is clear that humanities, social sciences, arts, and law all have important places as components of Indigenous studies. In Indigenous thought, however, it is important to keep in mind that none of these areas should be separate from the relationships between humans and the natural world. Indigenous knowledge and philosophy is inherently interdisciplinary because it links the human and the nonhuman. This link is the basis not only for Indigenous concepts of nature, but also for concepts of politics and ethics. In contrast to the Western tradition, in Indigenous thought there are no clearly defined boundaries between philosophy, history, sociology, art, law, biology, geology, anthropology, and geography.

It seems crucial that the natural world be considered an integral part of the study of Indigenous peoples and their cultural traditions and belief systems; however, the opposite appears to be true. Such departments and degree programs specifically or implicitly exclude disciplines that deal in any meaningful way with the natural world. A typical advertisement in academia for a faculty member in a department or program that focuses on Indigenous people invariably includes language such as the following: "We recognize that (fill in program or department title) is a new and interdisciplinary field. We welcome applications from individuals specializing in areas that might include American Indian Studies, Native American

Studies, *or in related fields such as anthropology, sociology, education, literature, economics, law, history, or political science*" (emphasis added). As an example, a major midwestern university was recently seeking a new faculty member in Native American Studies. Their advertisement for area of research specialization includes the following disciplines: "sociocultural anthropology, folklore, linguistic anthropology, ethnomusicology, Native prehistory, archaeology, oral literature, contemporary social and political life, or study of Native cultural traditions, but self-nominations from all fields are welcome. The successful appointee will have a tenure home in an appropriate department, possibly in Anthropology, Communication and Culture, or Folklore and Ethnomusicology." The areas of academia that are excluded from such consideration are also the areas of academia that are closest to traditional Indigenous knowledge and philosophy. Why are economics and anthropology, which are traditionally two of the most colonizing disciplines in academia, considered to be fields related to American Indian studies or Native American studies whereas biology, geology, or even geography are not?

As a consequence academic programs dealing with Indigenous peoples have essentially, albeit ironically, opted to create Indigenous studies in the best tradition of Western metaphysics, thereby incorporating and institutionalizing the colonized perspective at the same time that many such programs are supposedly attempting to address the issue of decolonization. As an example, the Indigenous Nations Studies Program at the University of Kansas provides the following mission statement: "The Center for Indigenous Nations Studies contributes to the *survival, self-sufficiency, mutual support, empowerment, and decolonization of Indigenous Peoples throughout the world.* The Center, working from a cross-cultural and global perspective, educates students through a combination of instruction, study, research, and practical experience" (emphasis added). All of the goals listed above would be enhanced by inclusion of the natural sciences into this and related programs.

ACADEMIC DISCIPLINES AND INDIGENOUS THOUGHT

This point can be emphasized through brief examination of the bases of the academic disciplines typically associated with programs in Indigenous studies as they exist in Western academic institutions. The humanities, as is implicit in the word itself, deal solely with humans and exclude the nonhuman world. The disciplines of English (literature), history, art history, and various non-Indigenous language-based departments (Germanic, Romance, Eastern European, etc.) have little or no relationship to Indigenous traditions or knowledge. In contrast, as I pointed out in Chapter 7, Indigenous literature is heavily influenced by themes from nature and ecology.

The social sciences have an even more curious historical tradition. Disciplines in this grouping all study the human species from a perspective considered to be scientific in its basic principles. Yet these disciplines tend to function as if the principles they develop had no application or relevance to the nonhuman world. Implicit in these approaches is the Western concept, derived from Bacon and Descartes following an intellectual tradition established by Aristotle, that the study of the nonhuman must be kept separate from the study of the human. The former is the province of the natural sciences, such as biology, and the latter is the subject of several subdisciplines, such as anthropology, psychology, and sociology.

A few social sciences dip cautious toes into nonhuman waters. For example, comparative psychology often employs rats or pigeons as subjects and simple models for mental processes. Physical anthropology can include nonhuman primates, but in general, anthropology, as is inherent in its name, deals solely with humans. Those social sciences that look at nonhumans assume that the nonhuman study subjects are little more than simple models for the more complex behavior of humans. In fact, one important criticism of anthropology by Indigenous peoples in many parts of the world is that those who practice anthropology often treat their study subjects (humans) as if they were less than human, or at least lesser than the anthropologists (Thomas 2000). Perhaps more importantly, recent work has shown that barriers between human and nonhuman are becoming blurred, and it is apparent that many nonhumans show behavior and cultural traditions once considered to be solely attributes of humans. Anthropology can be the butt of Indigenous humor; for example, describe a typical Indian family—answer: mother, father, two children, one grandparent, and an anthropologist. Despite this disdain on the part of Indigenous people, within academia it seems more appropriate for Indigenous scholars to obtain degrees in anthropology than in biology.

Fields of study such as creative arts and writing, as well as professions like law, business, and medicine, are also clearly human centered, although as discussed in Chapter 7, Indigenous artists and writers often emphasize relationships between humans and the nonhuman world. Medicine resembles psychology and anthropology in its dependence upon empirical studies involving nonhumans as the first steps in its investigation of the effects of various treatments and medicines targeted at humans. This remains fundamentally different, however, than studying nonhumans and their relationships with humans as a subject with its own merit.

In addition to these disciplines there exist several "applied" sciences such as geography, forestry, wildlife management, and engineering. These disciplines emphasize human relationships to the natural world; however, they treat various components of the nonhuman world (such as rivers, forests, and wildlife) either as (1) resources to be managed for human benefit, or (2) problems to be dealt with in an effort to minimize the discomfort and inconvenience experienced by humans. To be fair, these disciplines are also typically excluded from consideration for faculty appointments in Indigenous-themed programs and departments.

A truly Indigenous approach within academia should emphasize the links between these superficially disparate fields of study and the relationships that all have to the natural world. As emphasized above, the failure of Indigenous studies programs to incorporate the natural sciences as a component of their discipline may result from the lack of respect that has been shown to traditional knowledge by Western scientists. A problem often faced in trying to get Indigenous students to consider careers in the physical and natural sciences is overcoming the belief that "science is not for Indigenous people" because Western science has a long history of denigrating Indigenous knowledge and beliefs (Deloria 1995). Western culture often treats the knowledge base of Indigenous cultures as "primitive superstition," which is characterized as subjective and personal, as opposed to the "rational" perspective of Western scientific knowledge.

Such attitudes on the part of academics, both Indigenous and non-Indigenous, ignore the body of Indigenous knowledge relative to the functioning of natural systems, and especially to the science of ecology. As a result of such narrow biases, many Indigenous scholars and students do not think of modern science as a discipline relevant to their people. It is important to emphasize, however, that Indigenous peoples developed ecological concepts because their very existence depended on understanding ecology, especially relationships between species (connectedness) and the nature of individuality within species. One way to end the intellectual and academic marginalization of Indigenous studies may be to demonstrate that a truly Indigenous-based, and therefore more integrated, knowledge of the world may offer advantages in solving the problems of contemporary industrial societies.

Examination of Indigenous scholarship, creative writing, and art reveals that it is permeated with images from and descriptions of relationships with the natural world. Native American or American Indian studies programs will become more inclusive of Indigenous thought and will function more effectively if they reach out to Indigenous scientists and work to incorporate Indigenous perspectives on the natural sciences as part of their pedagogy. In Indigenous literature, connection to nature is seen as a way of restoring Indigenous identity and of healing individuals who have been damaged by their interaction with the dominant culture. In Chapter 7 I demonstrated that many major contemporary Native American novelists employ constant reference to nonhumans and to the natural world in establishing the nature of their characters and the links that these characters have to their communities. Place is very important in the identities of the Indigenous characters created by these authors. Examples include the authors discussed in Chapter 7, as well as Linda Hogan, Thomas King, Joseph Marshall III, Irvin Morris, Simon Ortiz, David Seals, Luci Tapahonso, and James Welch.

Perhaps the only prominent Indigenous writer who does not use nature and the links of Indian people to nature as a regular theme in his writing and seems to have little comprehension of how humans interact with nature is Sherman Alexie, who not surprisingly, may be the Indigenous writer most popular and widely accepted by non-Natives in America today.

In art most important Indigenous artists incorporate nature and the links between human and nonhuman as a regular theme, if not the principal theme, of their work. The sculpture and carvings of the great artists of the Pacific Northwest, such as the Haida Bill Reid, all deal with representations of nonhumans and their links to one another. Painters who come from peoples of the north woods, such as Anishinaabe Roy Thomas and Cree Norval Morriseau, also use the nonhuman and links to humans as the major themes of their work. In the Kiowa and Comanche cultural tradition, most of the important paintings of Blackbear Bosin portray nonhumans and their interactions with humans, a tradition being maintained in current times by Barthell Little Chief.

History is another discipline within the humanities where Indigenous links to nature are important. As mentioned above, human history existing independent of place and nature is a concept foreign to Indigenous peoples, because their history cannot be separated from the entire biology and geography of which they are a part. Historical accounts provided by Indigenous peoples emphasize this relationship to nature and the importance of specific localities. Many traditional stories, including creation stories, do not deal with the exact time when events happened; however, they are always very specific about the localities where these events occurred. Many of the events that are described happened so long ago that they exist "on the other side of memory" (Marshall 1995). As long as Indigenous people remain connected to the places where their cultures evolved, and their memories remain intact, their sense of history remains intact, even though the exact time when these events occurred is of little significance.

This sense of place illuminates an issue little understood by non-Indigenous geneticists, who persist in contending that they are helping Indigenous peoples to "find out where they have come from." To most Indigenous peoples this is a nonquestion—Indigenous cultures have arisen in the localities they identify as their sites of origin, which is why they are truly Indigenous. If more Indigenous people had degrees in the natural sciences, it would be much easier to establish dialogues with geneticists and explain both the costs and benefits of genetic research to Indigenous communities.

Philosophy is another discipline within the humanities where Indigenous links to the natural world are important. The essence of Indigenous philosophy is that one be native to a place and live with nature. In his discussion of "Indian Wisdom," the Lakota Luther Standing Bear stated:

> The Lakota was a true naturist—a lover of nature. He loved the earth and all things of the earth. . . . This is why [an] old Indian still sits on

> the earth instead of propping himself up and away from its life giving forces . . . to sit or lie upon the ground is to be able to think more deeply . . . he can see more clearly into the mysteries of life and come closer in kinship to other lives about him. . . . Kinship with all creatures of the earth, sky, and water was a real and active principle. . . . The animals had rights—the right of man's protection, the right to live, the right to multiply, the right to freedom, and the right to man's indebtedness—and in recognition of those rights the Lakota never enslaved the animal, and spared all life that was not needed for food and clothing. (Standing Bear 1978: 192–193)

The philosophical concepts outlined by Standing Bear contrast with those of the dominant Western (European) philosophy, which assumes that humans live above, separated from or in opposition to nature (Suzuki and Knudtson 1992; Martin 1999). Western philosophers from Aristotle to Descartes to Locke have emphasized separation of humans from the natural world (Mayr 1997; Pierotti and Wildcat 1997a, 1997b, 1999b, 2000, 2001; Coates 1998).

As laid out by Standing Bear and others, Indigenous philosophy is based on the concept that it is best to live with the geography and biology of your environment without trying to alter it solely to meet human needs, while casting off the modern Western view that "space" exists to be conquered (Cordova 2007). This philosophy has allowed Indigenous people to take their knowledge of the natural world with them, despite relocations, both forced and voluntary, and has allowed them to survive these experiences and establish sacred places in their new homes (Owens 1998: 164).

Within an Indigenous ethical system, nature exists on its own terms, and individual nonhumans have their own reasons for existence, independent of human interpretation. Living with nature requires people to rearrange the customs and habits of their daily life. One such custom involved representing sound ecological management in strongly ethical (or religious) terms and developing a view of the environment that stressed specific concrete bonds between nature and the human community. The cultural diversity of Native Americans reflects their intimate ties to the land and the biology of the places that they call home in specific social codes and institutions.

Indigenous philosophy encompasses both scientific knowledge and religion, in the sense that religion is the ritual representation of the community and a device for sanctioning moral and ethical codes. "The task of the tribal religion . . . is to determine the proper relationship that the people must have with other living beings" (Deloria 1992), which means that tribal religions encode and embody knowledge about the environment and human relationships with nonhumans. In this sense, culture, religious tradition, and scientific knowledge are intimately and pragmatically linked, demonstrating the interdisciplinary nature of Indigenous knowledge (Anderson 1996; Pierotti and Wildcat 1999b, 2000; Deloria and Wildcat 2001; Cordova 2007).

Given the above, it should be no surprise that knowledge about the environment and nonhumans is a major component of Indigenous education (Deloria 1990; Deloria and Wildcat 2001). In particular, Santa Clara Pueblo educator Gregory Cajete stresses that Indigenous education should be based upon ecological principles and knowledge. In his book *Look to the Mountain: An Ecology of Indigenous Education* (1994), Cajete systematically examines the differences between Indian and non-Indian understandings of how education should function and argues that Indigenous educational philosophy should be rooted in environmental knowledge and links to the natural world. A series of essays on education by Vine Deloria Jr. argues that Indigenous education must be rooted in relationships between Indigenous people and the natural world (Deloria 1990).

It should be clear from these arguments that the natural sciences and Indigenous scientists must be an important component of Indigenous or Native studies programs. There is no component of traditional Indigenous life that is not influenced by the natural world, and Native Americans would not be the peoples they are without their links to nature and the nonhuman elements of their social groups. The distractions and harm to traditional Indigenous ways of knowing that have resulted from numerous institutionalized attempts at assimilation are no excuse for allowing studies of Indigenous peoples and their traditions to be academic exercises that are merely grafted onto Western conceptualized and intellectually grounded academic disciplines.

The future of Indigenous peoples lies not in the greed- and fear-based concepts of the Renaissance and the "Enlightenment" of Western European tradition, which are likely to lead the human species to destruction. To truly be Indigenous, Native American scholars must begin studying links to the places where they have lived and currently live. Even more importantly, links must be re-established to places where Indian children and grandchildren will live, whether these are cities, small towns, reservations, or homelands, and must always include the geography and nonhuman inhabitants of these places.

There is much work to be done wresting American Indian, Native American, and Indigenous studies programs from the underlying philosophy and worldview to which the American university typically conforms. Let us seriously Indigenize such programs. A good step in this direction would be to find ways to effectively incorporate the study of nature and landscape into what must necessarily be an integrated or interdisciplinary field of study and restore understanding of the natural world.

To close I would like to address this last point by returning to the question posed in my title: "Who speaks for the Buffalo?" As I discussed in Chapter 7, one of the most powerful images ever generated on this topic is the description of a buffalo hunt in Louise Erdrich's 2001 novel *The Last Report on the Miracles at Little No Horse*. To me this story told by Erdrich illustrates the multidisciplinary aspect of Indigenous culture and thought (Chapter 7). This story incorporates both literature and history. This story is also used to indicate sociological and psychological facets of the Indigenous experience, and it clearly could be used in anthropology as well. It is crucial to remember, however, that this account describes the behavior of a group of nonhumans, but the behavior described is one that the Western scientific tradition is incapable of explaining. It is clear, however, that Indigenous people are capable of providing insight and meaning to an experience from which Western scientists would retreat in fear and frustration. Links between the Ojibwe and the buffalo allow understanding and even identification.

The answer to our question "Who speaks for the buffalo?" is that Indigenous people must. Louise Erdrich, an Indigenous novelist, speaks for the buffalo in a way that no non-Indigenous person, and certainly no Euro-American conservation biologist, could. Her speech comes from her acknowledgment that Indigenous Americans and the nonhuman are connected in deep and profound ways. Similarly the Kluane people in the Yukon of Canada speak for the nonhumans that they think determine their identity and way of life as a traditional people (Nadasdy 2003). By acknowledging links to the nonhuman world, we can develop Native American or Indigenous studies programs and departments that acknowledge Indigenous heritage and truly represent what it means to be Indigenous.

11 Traditional Ecological Knowledge

The Third Alternative

> Capitalism and communism are simply the opposite sides of the same Eurocentric coin. What the world needs is not a choice between capitalism and communism, between one aspect of eurocentrism or euro-supremacism and another. What we need is a genuine alternative to the European tradition as a whole.
>
> Russel Means (Lakota), quoted in Churchill (1995)

Over the last century there has been considerable debate over the appropriate way in which Americans, and other peoples, should treat the natural world (Leopold 1948; Dunlap 1988; A. Wilson 1992; N. Smith 1984, 1996). In the United States, Europe, and other "developed" countries there are a large number of people who advocate a pro-development extractive approach. In this approach natural resources are perceived largely in terms of their economic value to humans, and there is little or no concern about negative impacts upon nonhumans, either as individuals or as populations, except in a narrow economic sense. This perspective clearly dominated attitudes toward environmental issues and resource management until the 1960s (Dunlap 1988) and is currently exemplified by the "wise-use" movement (Lehr 1992) and most economists.

This extractive approach has attained new currency in the politics of neoconservative politicians and businessmen who wield considerable influence in modern political circles. This viewpoint has generally been identified with the political right (N. Smith 1984, 1996); however, exploitative approaches may come from all shades of the political spectrum, including the supposedly liberal. In the latter case, concerns about jobs and economic development are usually touted as a justification, even when there are severe negative impacts on environments and ecosystems.

There also exists a set of opposing models, stemming from the romantic philosophical tradition, which argue that nature and nonhuman animals must be protected from human interference. Such approaches assume that true conservation means setting aside tracts of land from which human settlements and even humans themselves may be excluded (Brinkerhoff Jackson 1994; Owens 1998). As a primary example, the U.S. Wilderness Act of 1964 defines wilderness as space forever untrammeled by man (Owens 1998) or as "a place where man himself is a visitor who does not remain" (Gomez-Pompa and Kaus 1992). This preservationist approach is set up as if it were in opposition to the extractive, economic driven approach, but advocates of preservation capitulate on many key issues—for example, they often tout ecotourism as a means of economic development over other supposedly more extractive forms, even though ecotourism can be very destructive to local habitats (e.g. coral reefs, deserts).

One example of the preservationist way of looking at the world is the extraordinary and artificial efforts that have been made to keep New Hampshire's "Old Man of the Mountain," a natural rock formation that bore a vague resemblance to a human profile, from undergoing natural erosion and slippage. Despite chains and bolts that were meant to hold it in place, this formation recently collapsed. In response, a number of "environmentalists" and "nature lovers" have argued that it should be artificially reconstructed so the "view can be restored," despite the fact that this is not a natural view, nor has it been so for many years.

The preservationist viewpoint has been identified with the political left (A. Wilson 1992; N. Smith 1984, 1996), but as with pro-development forces, "conservationists" may be represented throughout the political spectrum. Many conservatives want areas preserved that are of personal interest to them, such as hunting preserves or spectacular views.

An example of this approach can be seen by looking at the "17 Mile Drive" along the coast of the Monterey Peninsula in California. When I was an undergraduate at the University of California at Santa Cruz, I helped a UCSC graduate student who was surveying sea otters, *Enhydra lutris*. When California governor Ronald Reagan and his chief of California Fish and Game, Ray Arnett, were threatening to cave in to abalone fishermen who wanted sea otters controlled, this grad student started a petition drive to prevent interference with the recovery of the endangered otters. He went to Del Monte Country Club during the annual midwinter pro-am golf tournament and was able to obtain signatures from a large number of corporate CEOs and other major players in the California economy. The signatures on this petition were helpful in causing Governor Reagan to change his position. A decade later when my wife was working as a research diver assessing the impact of the otters on kelp forests in the same area, one of the Pebble Beach golf course members asked who the divers were and what they were doing. Country club staff told them that the divers were just "underwater gardeners, who took care of the kelp forest," which satisfied the club membership. It was a standing joke, albeit with serious overtones, that the reason sea otters were allowed to recover and even expand their range was because they were "cute" and had the good sense to live alongside some of the most expensive real estate in North America.

Despite superficial apparent differences, all Western European attitudes toward nature come from the same European philosophical roots—Descartes, Bacon, Locke, and the Enlightenment (N. Smith 1984, 1996; Coates 1998). These traditions assume that humans are autonomous from, as well as in control of, the natural world. As described in the writings of philosophers ranging from Aristotle and Descartes to Kant, humans are considered to be creatures apart from the rest of life (Mayr 1997; Cordova 2007). For my purposes, I assume the viewpoints discussed to be characteristic of the dominant cultures in most of the world where philosophical values derived from Western European intellectual traditions dominate economic and cultural thinking. In Europe, North America, and Japan the vast majority of citizens live in industrialized societies where nature is viewed as separate and "under control" (N. Smith 1984, 1996). Individual humans from across the political spectrum in these societies view the natural world as consisting of "resources," which carries the implicit assumption that all of nature can be exploited for human benefit, regardless of whether it is for economic or aesthetic purposes.

Citizens of industrialized societies typically adhere to the perspective that nature, like wilderness, truly exists only in places that are separate from humans (Leopold 1948; N. Smith 1996; Owens 1998). These attitudes fail to recognize that the problem is not that knowledge does not exist to effectively manage resources, but rather that the real problem is motivating humans to treat the natural world with appropriate respect (Anderson 1996; Pierotti and Wildcat 2000, 2001).

Some conservationists have contended that the approach they use is in the spirit of Native American or Indigenous traditions (Pierotti and Wildcat 1997a, 1997b, 2000, 2001). Such associations are based upon false assumptions about the true nature of Indigenous belief systems, because unlike Western philosophy, TEK assumes that humans are, and always will be, connected to the natural world and that there is no such thing as nature that exists independent of humans and their activities (Deloria 1992; Pierotti and Wildcat 1997b; Owens 1998).

Some connections that are a crucial aspect of TEK are also based on extraction, but in this case the extraction is done with respect and should not damage the resource. For example, animals or plants may be taken for food or for the manufacture of clothing or shelter, combined with recognition of the inherent value and good of nonhuman lives (Taylor 1986, 1992). TEK is based on the premise that humans should not view themselves as responsible for nature. We are not stewards of the natural world but instead are a part of that world, no greater than any other part (Standing Bear 1978; Pierotti and Wildcat 1997b). TEK deals primarily with motivating humans to show respect for nonhumans. The respect for the nonhuman inherent in TEK constrains natural human tendencies toward overexploitation, because nonhumans are incorporated into the ritual representation of the community and are considered as members of the community (Anderson 1996; Barsh 1997).

Numerous publications have discussed differences between the worldviews and knowledge base of Indigenous peoples and that of the "dominant" or "Western" culture (e.g., Johannes 1989; Mander 1991; Suzuki and Knudtson 1992; Anderson 1996; Hughes 1996). As discussed in Chapter 9, one major difference between Indigenous peoples of North America and Western European immigrants to North America is that people of Western European ancestry tend to assume that "progress" is a natural state of affairs, which constrains them to look backward and forward in time to get a sense of their place in history. Their moral instructions are in written form and at least in theory invariant, such as the Bible or the Quran. Despite repeated translations and attempts at reinterpretation, these instructions are a minimum of a thousand years old, which means that despite their relevance when they were written, they have not been changed to accommodate the massive social and technological changes of the intervening period of time. This leads to an odd sort of conjunction, although Western religious beliefs and Western science are often considered to be oppositional, both assume that once a scientific principle is laid out, it is set in stone and applicable to all cultures.

In contrast, Indigenous peoples look to the physical spaces in which they live to get a sense of their place in history and there is little discussion of the future in Indigenous thought (Deloria 1992; Basso 1996; Pierotti and Wildcat 2000). The stories told are related to specific locations in the immediate landscape, which allows those hearing the stories to identify readily with history. The idea of human history existing independent of local places and the natural world is foreign to the Indigenous peoples of North America, because for them their history cannot be separated from the entire geography, biology, and environment to which they belong. "In the traditional [way of knowing], there is no such thing as isolation from the rest of creation" (Deloria 1990: 17). One clear example of the process of thinking spatially can be seen in the Indigenous tradition of invoking the four horizontal directions (east, south, west, and north), as well as to the sky and the earth. People making such prayers acknowledge the space in which they live, and their understanding that the creative forces that shape their lives exist in the natural world that literally surrounds them in all of these directions.

If efforts are made to restore and preserve traditional knowledge and philosophy, TEK can serve as an intellectual foundation not simply for Indigenous science, but also for an Indigenous theory and practice of politics and ethics centered on natural places and connection to the natural world. These traditions are capable of generating a conservation ethic on the part of those who follow its principles, which includes 1) respect for nonhuman entities as individuals and the existence of bonds between humans and nonhumans, including incorporation of nonhumans into ethical codes of behavior, 2) the importance of local places, and 3) recognition of humans as part of the ecological system, rather than as separate from and defining that system. As I have argued previously, TEK is based upon empirical observations resulting from patient observation of the natural world and its patterns, and it is also inherently multidisciplinary because it links the human sphere with that of the nonhuman. TEK is not only the basis for Indigenous concepts of nature but also the basis for concepts of politics and ethics. Clearly defined boundaries do not exist between philosophy, history, sociology, biology, and anthropology in Indigenous thought (Pierotti and Wildcat 2000).

In essence, TEK requires one *to be native to a place* (Jackson 1994), and *to live with nature* (A. Wilson 1992), in contrast to the dominant Western worldview that assumes that "civilized" humans live above, separated, or in opposition to nature (Mander 1991; Suzuki and Knudtson 1992; Anderson 1996). To be native to a place is linked to the concept of Indigenous and means that it is essential to live with the geography and biology of your environment without trying to alter it solely to meet human needs. Thus, TEK is expressed in the ability to experience a sense of place while casting off the modern Western view that "space" or nature exists to be conquered.

One of the best examples of this perspective can be found by returning to our example of the Maya people of southern Mexico and Guatemala from Chapter 1: "The Mayas . . . created brilliant sculpture . . . and architecture. . . . About the year 900, *for reasons no one has ever been able to explain, the Mayan leaders abandoned their magnificent structures and the race fell into an inexplicable decline.* During the Classic period, experts estimate that there were more than 200 people per square mile. . . . Today you can divide that number by 100 (Cahill 1993: 169; emphasis added).

There is great irony in this series of statements. Cahill clearly intends to be complimentary; yet it is obvious that his criteria for a successful culture are for it to have large buildings and a high density of human beings. As pointed out in Chapter 1, a very different interpretation of the Maya can be found in contemporary studies by researchers who work to understand the Maya perspective:

> The Maya civilization, one of the greatest of the ancient cultures, is by no means dead. Millions of Maya Indians, speaking two dozen related languages still live in Central America. They practice traditional corn agriculture and maintain many pre-Columbian rituals. Yet they are no more "survivors" of the "past" than are modern Englishmen who still eat bread and beef and worship in the Church of England. Maya civilization is *dynamic, living, changing, and above all creative*. Tough and independent, its bearers have adapted to the modern world. . . . The Maya of central Quintana Roo and nearby parts of the peninsula successfully rebelled against Mexico in the mid-19th century and remained completely independent until well into the 20th . . . fascinated with the outside world, yet determined to hold their own against it." (Anderson 1996, 74: emphasis added)

These conflicting views reveal the naïveté in many Western views of both Indigenous peoples, and even of what civilization means. It rarely occurs to people of European ancestry that the Maya might have chosen to abandon their cities and the large-scale trappings of their "civilization" because they were ecologically destructive and unfeasible. Today there are as many Mayans as there ever have been, but that they now choose to live at low densities in a more traditional and superficially "primitive" manner (Anderson 1996). Mayans are as sophisticated and intelligent as ever, yet they choose to minimize their impact on their immediate environment. Are the Maya functionally different than wealthy Americans of European ancestry who choose to live "in the country" once they can afford not to live in cities? Why is such a decision by Indigenous people evidence of *an inexplicable decline*, yet in people of European ancestry it is considered to be a mark of success to be able to spend your life in rural areas, supposedly *"getting back to nature"*?

This example demonstrates that Indigenous people are very different than the comfortable and romantic image of Rousseauian "noble savages." They are real beings who live in a reality where understanding the natural world is essential to survival, both of individuals and the long-term survival of the culture itself. Living with nature bears little relationship to such concepts as "love of nature," "closeness to nature," "communing with nature," or "conservation of nature," which are statements made by Western conservationists, often in reference to Indigenous peoples (see below and also Anderson 1996; Krech 1999a). People who feel that it is within their power to conserve nature also feel that they are in control of nature, and that nature should be conserved only insofar as it benefits humans, either economically or spiritually (N. Smith 1984). Within a TEK-based ethical system, nature exists on its own terms, and individual nonhumans have their own reasons for existence, independent of human interpretation.

LACK OF AN IMMIGRANT EXPERIENCE

One important thing to keep in mind is that Indigenous peoples lack an immigrant experience within their memories. I discussed the scientific aspects of this point in Chapter 9. All I wish to do here is emphasize two points: First, it is clear that the peoples of the Americas share ancestry with peoples of Asia, although the exact timing of dispersal events is unlikely to ever be clearly established. Second, regardless of whether humans have been in the Americas for only 15,000 years or as long as 25,000 years (Goebel et al. 2008), what is almost never emphasized is that even 15,000 years bp represents a time many thousands of years before any of the cultures of the modern world existed. None of the European or Asian nations from which people emigrated to post-Columbian America would exist for at least 10,000 years; in fact, Europe itself had not had modern human (Cro Magnon) residents for long when humans first came to the Americas (Balter 2001; Haak et al. 2005).

This is what it means for the origins of the Indigenous peoples of the Americas to exist "on the other side of memory" (J. Marshall 1995: 207).

What negates any alleged "immigrant" experience is that the worldviews and cultures of Native American peoples evolved in the environments of the continents of North and South America. Indigenous peoples depended upon the animals and plants of these environments for food, clothing, shelter, and, what is important but rarely discussed, companionship. In consequence, peoples developed strong ties to these nonhuman lives. "Little emphasized, but equally as important for the formation of [Native] personality was the group of other forms of life which had come down over the centuries as part of the larger family" (Deloria 1990: 16–17). As these places and beings existed and changed along with them for thousands of years, Indigenous peoples developed their cultural senses of place, along with their flexible knowledge base. These values have been kept intact through TEK, regardless of whether the people have been forced off their original lands, either by changing ecological conditions or by European immigrants (see Chapter 10 in Owens 1998).

Living with nature requires people to rearrange the customs and habits of their daily life. As outlined in Chapters 2 and 3, the origins of TEK are based on the awareness that Indigenous societies existed under conditions of constant fluctuation in the resources upon which they depended. Given the constant possibility of shortages, a means had to be found to convince communities and families to economize with regard to their use of natural resources (Anderson 1996; Henriksen 2009), in fact it has been argued that the constant threat of starvation has been one of the "most powerful and cohesive forces in Native life" (Ross 1992: 133–139) and drove the development of ethical systems based on placatory interactions with the natural world.

Such an ethical system almost certainly arose from early experiences of Indigenous Americans. Whatever the cause, many species of large mammals disappeared after humans arrived in the Americas, some of which were probably important sources of food and materials from which clothing and shelter could be manufactured. Regardless of whether or not Indigenous peoples were responsible for the extinction of these megafauna as argued by Paul Martin (2005), or whether these organisms went extinct as result of major environmental changes (Guthrie 1990; Grayson 1991), or as has been recently suggested, Indigenous people arrived well before the actual extinction events, but still had a role in initial declines in megafaunal populations (Gill et al. 2009; Johnson 2009) it is clear that Indigenous Americans were contemporaneous with these organisms.

Stories exist from the traditions of some Indigenous peoples that seem to refer to megafauna. Part of the Cheyenne creation story refers to a time when there were "two types of bison. Those whose horns turned upwards like the bison of today and *those whose horns pointed downward*" (H. Mann, personal communication, emphasis added). If we consider large shaggy mammals who have hornlike structures that point down, I suspect that this story refers to either mammoths (*Mammutus*) or mastodons

(*Mammut*), both of which have large tusks that in fact point downward. According to Dr. Mann, the "bison whose horns turned downward" were larger and more dangerous than the bison of today, which also supports the idea that these stories may represent early memories of proboscideans and also suggesting that they would have been unlikely targets of human predation. The Abenaki peoples of the northeast have stories that appear to refer to megafauna, including mammoths, sabre-toothed cats (*Smilodon*), and great bears (J. Bruchac, personal communication).

If we assume that the early Indigenous Americans had philosophical and spiritual traditions similar to those of their descendants at first contact with Europeans, they almost certainly had "relationships" with the megafauna similar to those they maintained with their nonhuman relatives both prior to and after contact with Europeans (Barsh 2000). Some of these megafauna were almost certainly regarded as powerful entities, such as mammoths and mastodons, sabre-toothed cats, short-faced bears (*Arctodus*), cave lions (*Panthera atrox*), dire wolves (*Canis dirus*), ground sloths (*Megalonyx* and other genera), and giant beaver (*Castoroides*) (see P. Martin 2005, Table 2 for a list of North American megafauna). The disappearance of these powerful contemporaries may well have been one factor that convinced the First Americans that nature was not predictable and constant and that some key elements of their communities could in fact disappear forever.

I have already discussed the more recent example of salmon populations in the Pacific Northwest, which over the last several hundred years were highly variable and may have even disappeared for periods of time up to decades in length. This led to the tradition in which the salmon are treated today with respect because "there was a real and proven likelihood that they might not come back again in sufficient numbers to feed the tribe." Such experiences led to an oral tradition, which serves as "an epic parable about the capricious nature of nature" (Raban 1999: 404).

Once again, I rely on Indigenous literature to illustrate the point I am trying to make:

> The purple asters are growing in wide fields around the red rocks past Mesita clear to the Sedillo Grant. This year there has been more rain here than I have ever seen. Yesterday at Dripping Springs I saw a blue flower I had never seen before . . . So many of these plants had never bloomed in my lifetime and so I had assumed that these plants did not bloom; now I find that through all the years they were simply waiting for enough rain.
>
> I remember the stories they used to tell us about places that were meadows full of flowers or about canyons that had wide clear streams. I remember our amazement at these stories of lush grass and running water because the places they spoke of were all changed; the places they spoke of were dry and covered with tumbleweeds and all that was left

> of the streams were deep arroyos. But I understand now, I will remember this September like they remembered the meadows and streams; I will talk about the yellow beeweed solid on all the hills, and maybe my grandchildren will also be amazed and wonder what has become of the fields of wild asters and all the little toads that sang in the evening. Maybe after they listen to me talking about this rainy, lush September they will walk over the sandrock at the old house at Dripping Springs trying to imagine the pools of rainwater and the pollywogs of this year. (Silko 1981: 170)

In this short essay from her book *Storyteller* the Laguna Pueblo writer Leslie Silko manages to capture many of the themes I address in this book: the multigenerational nature of Indigenous knowledge and experience, the knowledge of how changeable the environment can be over long periods (decades), how stories can communicate these themes and make them truly traditional, and how these stories emerge from long-term experience of the nature in a particular place. Silko makes these stories personal and emotional, revealing how an individual Indigenous woman relates to her family across several generations, and how this links to the changeable but particular place they all call home.

It is likely that Indigenous Americans were constantly aware that the resources upon which they depended could disappear forever (see also Chapter 9 in Ross 1992). They tried to reduce the chances of this happening by developing an ethically based system of restraint with regard to hunting and use of other key resources (Deloria 1990; Rappaport 1971; Anderson 1996). The cultural diversity of Indigenous Americans reflects intimate ties to the land and the biology of the places that they call home in specific social codes and institutions, rather than in some misty "union with nature" (Anderson 1996). Thus, TEK represents both science and religion, in the sense that religion is the ritual representation of the community and a device for sanctioning moral and ethical codes (Durkheim 1961). "The task of the tribal religion . . . is to determine the proper relationship that the people must have with other living beings" (Deloria 1992). As mentioned in Chapter 1, in TEK, religion is used to teach environmental knowledge; therefore, it is not surprising that TEK is based on and has considerable insight into the workings of nature, and in many ways converges closely upon the Western science of ecology. The reason I refer to TEK as the *third alternative* is that, unlike with both the extractionist and conservationist points of view, it in no way assumes that humans are in control of nature; instead it requires that humans must live with nature if we are to survive.

This is why Krech (1999a, 1999b) was so wrong about assuming that Indigenous people were the equivalent of "environmentalists" or "conservationists" who draw their ideas and inspiration from the European "romantic tradition." It is very different to assume that you can control and define nature than it is to realize that nature will do what it will and that the best humans can hope for is an accommodation with forces beyond our control, or even our ability to imagine.

As mentioned previously, a major theme of TEK is that all things are connected. This theme of connection is not simply a homily or a romanticized cliché. Instead this idea represents a realization that no single organism can exist without the web of other life forms that surround it and make its existence possible. This concept is closely related to the Western discipline of community ecology, but as Barsh (2000) points out, it is based on observation of behavior and the assumption of social interactions between different types of nonhumans as well as between humans and nonhumans. As with the Western concept of community ecology, TEK places emphasis on inter-relationships between different species and individuals, and describes these interactions by employing the metaphor of a web. Connectedness also shares themes with the ecological concept of nutrient cycles (Pierotti and Wildcat 1997b). Although the idea of cycles, or circles, of life, is an integral part of Indigenous spiritual beliefs, these are not mystical concepts, but a practical recognition of the fact that all living things are literally connected to one another. As a result of these connections with the nonhuman world, Indigenous peoples do not think of nature as "wilderness," but as home. Natives do not leave their "house" to "go into nature," but instead feel that when they leave their shelter and encounter nonhumans and natural physical features that they are just moving into other parts of their home (Reichel-Dolmatoff 1996: 8–9).

In its root structure, Western ecology employs a similar concept. The word *ecology* comes from combining the concepts of *oikos* and *logos*, the Greek words for "house" and "knowledge" (Kingsland 1991). Thus ecology as a science has a specific definitional relationship to Indigenous concepts of space-based ecology in that it acknowledges that ecology is the study of the local space where an individual or community of individuals lives, the space considered to be their "home."

The shared ideas of connectedness and nature as home have profound implications for conceptions of politics and ethics. In Western political and ethical paradigms an understanding of how human beings ought to act ethically is imbedded in one's social, or human, relationships, and a "home" is perceived of as being located indoors. In contrast, Indigenous peoples see relationships among species and incorporate themselves within a community consisting of many nonhuman persons: four-leggeds, winged-ones, plants, and even landforms, such as mountains, rivers, hills, and streams (Deloria 1990, 1992; Pierotti and Wildcat 1997b, 2000, 2001).

Inclusion of other living beings and natural objects into the category of "persons," requires politics and ethics that include these other community members. Consideration of nonhuman entities that comprise ecosystems, including landforms, plants, and animals as individual persons who are part of their communities, operates to keep humans attending to the specific entity and its particular value. This emphasis on individuality provides a spiritual alternative to overgeneralizing about nonhumans (Anderson 1996). No Indigenous person will ever make a statement as naïve and cynical as a remark made by former California governor Ronald Reagan: "If you've seen one redwood tree, you've seen them all." Throughout Indigenous American cultures, there is a broad commonality of beliefs about animals in which human and nonhuman are bonded closely and part of one community, involved with one another in terms of empowerment and emotional interactions (Anderson 1996).

It is frustrating to Indigenous Americans to hear people of European ancestry speak romantically of Indigenous "closeness to nature" or "love of nature." This relationship is more profound than most people can imagine, and the implications of this relationship carry uncomfortable consequences. In nearly all Indigenous stories animal-persons and plant-persons existed before human-persons. Thus, these kin exist as elders, and much like human elders, they function as teachers and as respected members of our community. Acknowledging nonhumans as teachers and elders requires that we pay careful attention to their lives. This leads us to recognize that these lives have meaning on their own terms.

Awareness of the value of nonhuman lives extends the social world to include animals as well as humans (Barsh 2000). This is different than the Euro-American "conservationist" or "environmentalist" perspective. Humans are assumed to exist in mutual aid relationships with the nonhumans. If humans eat or otherwise use nonhumans, they are empowered by that relationship, which leads to mutual respect (Anderson 1996).

Recognition that nonhuman lives have value, but that nature can behave in ways that can be harmful to individuals, requires some rethinking of how humans view the world in which we exist. We can caricature this as "humans *care*, nature does not." Humans, regardless of ethnic or racial origin, form attachments that involve our emotions. The objects of our attachments are not only our fellow humans, but nonhumans as well. As a result, we place great emphasis on the lives of individual organisms, especially those with whom we have such attachments, a situation that is clearly seen in human relationships with dogs—domestic wolves whose family structure is similar to our own. If bonds between individuals of the same or different species help these individuals to survive or reproduce effectively, this is a useful outcome. At times these bonds with other beings are not just emotionally satisfying; they also help humans to survive (Marshall-Thomas 1994; Mitchell et al. 1997; Bekoff 2006, 2007).

Humans link these bonds or attachments to "caring" and often extend this to "ethics." In Indigenous traditions this led to the idea of relationships among all living things (Barsh 2000; Pierotti and Wildcat 2000). These assumed relationships also lead to gratitude for the sacrifice of life when one creature gives up its life so that another may survive. As I have emphasized, Indigenous peoples were very aware of the changeable stochastic aspects of the natural world (nature may not *care*, but it can provide). Their solution was to try to convince nature and key food animals and plants that the humans were respectful, and as such, they hoped that these other beings would not take away their largesse.

In the Western (European) tradition, these experiences led to a different path, especially after the Black Plague (Pierotti 2004). This path assumed that, to survive, humans had to take charge of nature and control it in ways that reduced the impact of its unpredictable changes. These traditions had already started out along this path through the domestication of non-humans. By creating domestic animals Europeans and Asians had forever altered their relationships with nature (Manning 2004). This buffered these peoples against environmental fluctuations in food supplies, but ironically, this also created the environment that led to animal diseases making the jump to humans, leading to the plagues that decimated their populations. Experiencing plague convinced Europeans that even more control over nature was needed (Coates 1998; Pierotti 2004). This led to improved sanitation, dams, central heat and air, medicine, and so on, all of which create control, at least over microenvironments, but it also reinforced the idea that humans could be separated from nature.

This attitude came at considerable cost. Assuming "control" over nature led to a loss of the perception of relatedness and relationships with other forms of life among most people. Even a semblance of such a relationship would not be recovered until science took tentative steps in this direction starting in the 1950s and 1960s with the development of the fields of ecology and animal behavior. Ironically, these two disciplines are currently being marginalized and are often under attack in academic biology departments. In 2007 less than half of the biology departments in the United States required a course in ecology for their students majoring in biology, and less than 15 percent of these departments even offer a course in animal behavior (Cheesman et al. 2007). This is yet another reason that Indigenous people might step forward and reclaim these most Indigenous of scientific disciplines.

Attempts to control nature led to a philosophical tradition in which the "creator" imagined by such peoples must think (and look) as they do and that this "creator" must also require control over natural phenomena. This has led to arrogance in how scientists and resource managers deal with Indigenous peoples. As one Western scientist has put it: "Imagine people who confidently assume they can best describe and manage the natural resources of an unfamiliar region alone—ignoring local hunters, who know every cave and waterhole and the movements and behavior of a host of local species. Such, historically, has been the custom of most scientists and natural resource managers working in unfamiliar environments" (Johannes 1989: 5).

In the view of traditional Indigenous people and sophisticated thinkers from other traditions, the "creator" allows individual entities (both individuals and species or cultures) great freedom; however, they are constrained by what used to be called "natural law" (Davies 1994: 77–78), such as gravity, natural selection, and genetic heritage. Under this scheme these entities are continually presented with an environment that requires them to employ this freedom in creative ways. If the entities make a mess of things, the "creator" will allow them, or at least something like them, to try again (Davies 1994: 78), but it will not step in and help them.

A Lakota novelist described a similar line of thinking as follows: "The Christian God has a big lantern with the kerosene turned way up, and the people pray to him for help, for guidance and He lights the way. Now Wakan Tanka, when you cry to him for help, says, 'OK, here's how you start a fire.' And then you have to make your own torch" (Power 1994: 285–286).

Sometimes the variation that nature throws at these entities is beyond any control. Despite the best efforts of humans (whether spiritual or technological), weather, earthquakes, volcanoes, or collisions with asteroids remain beyond "control" of any kind. When such events happen, our inclination is to call the event a "natural disaster." If this disaster leads to considerable loss of human life, humans invariably ask "why?" The rest of nature, which does not assume control, simply responds, "why not?" and goes on trying to survive.

One thing I always find perplexing is that "disasters" are defined almost solely in human terms. I had an argument with a member of my PhD thesis committee, who was also trained in animal behavior and ecology, when he expressed relief that an oil spill had not come into a port city, but had instead "washed out to sea." I contended that humans were the ones responsible for the oil spill, and that they should be the ones to deal with its consequences, rather than the fish, seabirds, marine mammals, and invertebrates that were likely to be its victims at sea. It was apparent to me that he regarded the activity of the oil spill at sea as taking place in a figurative vacuum. This attitude is ethically indistinguishable from those of European immigrants who came to America thinking they were going to an "empty land." There is a large part of the Western philosophical tradition that assumes that anything that does not impact humans does not really happen, as in the old philosophical chestnut, "If a tree falls in the forest and no one hears it, does it make a sound?" No Indigenous person would ever ask such a foolish question; they would realize that even if no human ears were present, there would be thousands of other ears that would hear. Such thinking is a part of the Aristotelian legacy described in Chapters 1 and 2, but, and this is rarely discussed, it is also linked to the idea of wilderness as a place from which humans are absent (Gomez-Pompa and Kaus 1992).

These other beings are the children of Mother Earth as surely as are humans. Those that survive a disaster keep on eating, reproducing, and living their lives. Those that do not survive are missed, but life goes on. For example, when beaver were trapped out and exterminated in much of North America, the landscape and ecology was greatly altered, but beavers persisted, as did almost all of the species of North America, which adjusted to an environment in which beavers were now absent from many areas they had previously occupied. In those areas where humans have changed such behavior and stopped persecuting beavers, their populations have built up, other animals and plants have accommodated to the changes they have made in the landscape, and ecosystems have changed back in some ways to resemble their previous form. Still, this does not imply a balance of nature, for things will never be as they once were, but environments can change back to a semblance of their former selves if the right environmental conditions exist.

Accepting the fact that nature does not "care" in a human sense, that the environment fluctuates in uncontrollable ways, and that plants and animals continue to live their lives despite "disasters" presents a different way of viewing the world than the romanticized or sentimentalized Western perspective of a "balanced" nature, or even the idea of "North American Indians" "living in harmony with Mother Earth" as described by Hughes (1996) and caricatured by Krech and his followers. When populations plummet because of natural or manmade "disasters," individuals go on to live their lives and build up their populations. Many species that have been considered endangered, like the sea otters I discussed at the opening of this chapter, managed to recover when humans treated them respectfully. Today the nonhumans of North America seem to be waiting for their human relatives to return to their senses and learn to live side by side with them once again, in a manner that was only rarely peaceful but allowed reasonable numbers of all species to survive.

Once again I draw upon Indigenous literature to reinforce this point. The reason I do this is that there are few Indigenous ecologists, but Indigenous writers often illustrate important philosophical principles through their stories and essays. In this chapter I have already quoted Leslie Silko and Susan Power. Describing an encounter between a wolf and an Anishinaabeg man, the Anishinaabeg Louise Erdrich wrote the following:

> I spoke to the wolf, asking my own question: "Wolf" I said, "Your people are hunted from the air and poisoned on the ground and killed on sight . . . and stuffed in cages and almost wiped out. How is it that you go on living with such sorrow? How do you go on without turning around and destroying yourselves, as so many of us Anishinaabeg have done under similar circumstances?"
>
> And the wolf answered, not in words, but with a continuation of his stare. "We live because we live." He did not ask questions. He did not give reasons. And I understood him then. Wolves accept the life they are given. They do not look around them and wish for a different life, or shorten their lives resenting the humans, or even fear them any more than is appropriate. They are efficient. They deal with what they encounter and then go on. Minute by minute. One day to the next. (Erdrich 2005: 120–121)

To conclude, TEK is a constantly evolving way of thinking about the world. Although views covered by TEK are described as "traditional," this should not be taken to mean that they cannot change. The essence of traditional beliefs is that they have existed long enough for long-range consequences to affect them (Anderson 1996). As discussed in Chapter 1, use of the term *traditional* implies the repetition of a fixed body of data to many who are not of Indigenous ancestry (e.g., Berkes 1999). Each generation, however, makes observations, compares their experiences with what they have been taught, and conducts experiments to test the reliability of their knowledge (Barsh 1997, 2000). TEK is linked to both the long-range consequences of human action and environmental change; therefore adherents to TEK should always be able to modify their activities and responses if environmental conditions so demand and the proof of their effectiveness is their own survival (see also Alessa 2009).

This reliance on new information as local conditions change reinforces the spatial orientation of TEK (Deloria 1992; Pierotti and Wildcat 2000). The spatial orientation of Indigenous peoples leads them to recognize that there are always new experiences and knowledge in the world, and transmission of TEK by oral traditions allows them to adjust in response to changing conditions. As a result, ethical and moral instructions for living are fit to the current ecological and historical context.

It is important to emphasize this ability to change the way in which an individual or culture understands the world and its relationship to the world. This is true not only for Indigenous people but for people from any cultural tradition. There is nothing magical or "in the blood" about the ability to take the lives of other species seriously and to accord them respect and the status of "persons" (Pierotti and Wildcat 2000). An example of how this is being attempted by Western philosophers can be found in the work of Paul Taylor (1986, 1992) and C. D. Stone (1974), who explore the history of the expansion of the concept of legal "personhood" in the United States to include women, children, and nonwhites, the development of animal welfare laws, and the Western legal tradition of providing legal standing to such nonhuman entities as ships and corporations to create a basis for including nonhumans and features of the natural landscape in a philosophical and legal concept of "personhood." These philosophers draw solely on Western sources for their arguments, yet they are achieving a framework that creates a more respectful relationship between humans and the natural world.

Humans from any cultural tradition can establish a respectful and reciprocal relationship with nature if they are willing to be patient and put in the hard work. There are numerous examples where people of European ancestry managed to establish or at least maintain such relationships with nonhumans. In recent years several animal behaviorists have published books dealing with the emotional states and relationships of our nonhuman relatives (Marshall-Thomas 1994; Mitchell et al. 1997; Bekoff 2000, 2002, 2006; de Waal 1995, 2001, 2009; Fragazy and Perry 2003).

As an example, a PBS documentary with the unfortunate title "Killers in Eden" (shown 6 October 2005) discussed a mutualistic hunting relationship between Europeans and orcas (also called killer whales) off southeast Australia. This tradition was apparently originally established between Australian Indigenous peoples and orcas, and apparently the interaction was initiated by the orcas themselves, when they realized that humans could more effectively kill large whales than the orcas could on their own (Clode 2002).

In this situation the human settlement was located on a cape where large whales, especially humpbacks, were forced to pass through a narrow channel during their breeding migrations. When the orcas found a vulnerable humpback they would call the humans to help them by coming close inshore and slapping the water with their tails. This succeeded in drawing the attention of humans, even when it happened in the middle of the night. The two species hunted together based on what was called the "Law of the Tongue," in which the humans allowed the orcas to eat the tongues of the large whales. Since humans do not use whale tongues, but instead use blubber, bone, and sometimes meat, this situation worked out well for all concerned. Similar accounts of cooperative hunting have been recorded between orcas and salmon fishermen in British Columbia (Iglauer 1988). In a sad irony, both of these relationships were changed by betrayals of trust by humans who killed or shot at orcas, thereby breaking the relationship.

In a similar interaction between lions and Indigenous people in southwestern Africa, the relationship was changed by the removal of humans from an ecosystem where they had probably co-existed with lions for as long as 100,000 years. Within 10 years (two lion generations) lions changed from having a careful and respectful relationship with humans to an aggressive one (Marshall-Thomas 1994). These examples reveal two things: (1) how fragile these relationships are and how easily they can be destroyed, and (2) how the idea of "wilderness" as conceived of by "preservationists" of European ancestry can actually destroy relationships between humans and nonhumans.

The above accounts are similar to accounts of hunting between wolves and humans in North America, or to hunting associations between Indigenous Australians and dingoes, a form of domesticated wolf that left humans and returned to wild ways on the continent of Australia. In this case the humans raised dingo pups and then released them, knowing that the socialized dingoes would continue to help them locate and capture prey (Meggitt 1965).

Humans can establish relationships with other species; all it takes is respect and caring and living as though the lives of other species matter. Humans from a wide range of cultures already do this with wolves (domestic dogs), European and North African wildcats (domestic cats), horses, Indian elephants, and a wide variety of other species of birds and mammals. These are what we call pets or "domestic," but they are better described as "companion animals." It takes more work to establish such relationships with nondomesticated species, but it can be done. Barsh (n.d.) describes a wide variety of relationships between humans and nonhumans. But these relationships must be based on respect, not control or domination, in order for them to lead to a relationship with the natural world that is similar to that found in TEK.

I have tried in this book to show the reader that the knowledge, metaphors, and models employed by Indigenous peoples are as valid as those employed by peoples whose ancestry is in the nations of Europe or Asia. This should not be construed as an argument that Indigenous knowledge and models should stand alone, that Western society should co-opt Indigenous belief systems, or that Indigenous people cannot benefit in some ways from working with scientists from other intellectual traditions. These different ways of seeing and knowing can complement one another and in many cases advance human understanding. In some cases, however, Indigenous concepts have clearly had precedence and should be taken seriously by Western scientists. Recent discoveries of culture, traditions, and teaching by nonhumans (Bekoff 2000, 2002, 2006; de Waal 1995, 2001, 2009; Fragazy and Perry 2003) have shown that Indigenous peoples were correct in describing such phenomena. Similarly the social relationship-based approach toward ecology taken by Indigenous peoples continually yields new understandings of the dynamics of ecological communities.

An interesting thing about science is that, like TEK, it constantly changes and can ultimately be self-correcting. The problem with science as it is currently practiced is that too often it suffers from what I call the *Imo problem.* Imo was a Japanese macaque and apparently a genius for her species. When Imo was still a juvenile she discovered that if she washed pieces of potato she could get dirt off them so they were easier to eat. She later discovered that if she took a mixed handful of seeds and sand and threw them into water that the sand would sink and the seeds would float. Over several months she taught her mother, her playmates, and other adult females how to do these techniques (Itani and Nishimura 1973; Watanbe 1994; de Waal 2001). Rarely discussed in reference to this work was the discovery that there was one group of individuals who refused to learn from Imo, who was, after all, both female and a youngster. The group that chose to remain ignorant was made up of the old males in her troop. As Imo and her playmates grew up, her young male playmates learned from her teaching, but the old males never did. They went to their graves with grit in their teeth. Science is like Imo's troop in many ways. The old guard is most resistant to change, as in the case of Paul Martin and Vance Haynes, but also in the case of Vine Deloria Jr.

Fortunately we now have new generations of scientists who are open to the ideas of culture, traditions, high intelligence, and organisms establishing relationships that turn them into new types of life forms. These individuals are more open to TEK and different ways of thinking about the world. It is time for Indigenous people to overcome their uneasiness about Western science and realize that it is rapidly converging on TEK in many ways. It would be helpful for Indigenous people of all ages to make their contribution to overall understanding of the natural world and Indigenize science.

There is a powerful statement in the book *Neither Wolf nor Dog* about the potential that exists in relationships between Euro-Americans and Indigenous Americans that places this in context and also brings us back in a circle to the epigraph for Chapter 1.

> White people are jealous of us. If it hadn't been for your religion you would have lived just like us from the first minute you got to this land. You knew we were right. You started wearing our clothes. You started eating our food. You learned how to hunt like us. When you fought the English you even fought like us.
>
> You came to this country because you really wanted to be like us. But, when you got here you got scared and tried to build the same cages you had run away from. If you had listened to us instead of trying to convert us and kill us, what a country this would be. (Nerburn 1994: 137)

It is time for Indigenous people to acknowledge the power of their traditions and ways of life. To do this they will also need to reject the fear and narrowness that comes from religious traditions that have been imposed upon them for centuries. They will need to overcome the conundrum that trapped Vine Deloria Jr. near the end of his life. Deloria knew that Christianity was deeply flawed; yet he fell back on Christianity rather than Indigenous thought to provide answers when he saw flaws in Western science. He assumed that the problem was in the science itself, rather than in the narrow old men who had created the errors he saw. The solution will come, not from mockery, nor from falling back on Western religion to rebut Western science, but from merging Indigenous knowledge and perspectives on the natural world with those from science.

By placing TEK-based worldviews into a broad-based system of knowledge we will gain the ability to access a large amount of information and experience that has been previously ignored or treated as mysticism. This additional knowledge, with its empirically derived emphasis on the natural world, can provide scientifically testable new insights into some of the most pressing problems facing humankind today. The interdisciplinary structure inherent in TEK will make it relatively easy for knowledge and insights gained through TEK to be communicated among members of different disciplines, and can lead various stakeholders to negotiate more effectively with one another through a shared conceptual framework.

If we can successfully indigenize science, and through science indigenize all knowledge systems as embodied in academia, we will have achieved the end point described in the quote from Nerburn (1994) above. We will gain tremendous understanding and be able to provide real traditional leadership to a troubled world. What a people we will be and what knowledge and understanding we will have.

AFTERWORD

The approach I have described in this book has focused on my research fields, which are ecology, evolution, and animal behavior. However, Indigenous knowledge systems may have a great deal to contribute to other fields of biology, even those that have been historically the most reductionist in their approach. My Dine' colleague Wilfred Denetclaw, one of the first Native Americans to become a tenured professor in cellular and molecular biology, employs TEK-based approaches and stories to gain insights into the epidemiology of disease organisms such as Hanta virus (Denetclaw and Denetclaw 1994a, 1994b).

One of the most interesting discoveries where I think an Indigenous perspective can be merged with a distinctively Western approach to research is in the understanding of the origins of eukaryotic cells. As I have pointed out in other chapters, another concept recently rediscovered by Western science is that living creatures other than humans can alter their environments and even themselves in response to environmental changes (Jones 2005; Gerhart and Kirschner 1997; Kirschner and Gerhart 2005; Wright and Jones 2006). One example of such an interaction almost certainly happened with the proto-eukaryotic cells that became the ancestors of all multicellular life on earth. The formation of such cells involved a symbiotic merger of two types of prokaryotic bacteria through what is called *endosymbiosis* (Margulis 1998).

A new type of cell resulted from this combination of two different types of bacteria: One capable of metabolizing oxygen (aerobic respiration), which merged with larger cells capable only of fermentation (anaerobic respiration), joining and establishing a relationship. The aerobic form took up life within the body of the anaerobe and evolved into what we know today as mitochondria. Mitochondria are commonly thought of as organelles inside of eukaryotic cells (such as cells in the human body), but in actuality they are distinct organisms with their own genetic heritage that live symbiotically inside other cells. Mitochondria provide energy to their host cells through aerobic respiration. They are often metaphorically characterized as the "powerhouse" of the cell because they provide energy to the entire cell in exchange for being able to live in a nutrient-rich and protective environment provided by their hosts (Margulis 1998). These two types of interacting cells (mitochondria and their eukaryotic hosts) shaped each other. Without this relationship there would be no plants, animals, or fungi (Margulis 1998).

One of the most indisputable proofs of the relationship of humans to the rest of biological creation is that, like every other multicellular life form, we also contain mitochondria within our cells that are genetically distinct from "us" (since we think of ourselves as consisting of the eukaryotic cells that make up our physical selves and our genes as coming from only within our nuclei). No evidence of relatedness and interdependence is more compelling than the fact that humans share with the rest of eukaryotic organisms (animals, plants, fungi, protists, algae) a co-evolved relationship with our own mitochondria.

A recent book on evolutionary ideas (Prothero 2007) discusses endosymbiosis under the heading "Communal Living Builds Complex Cells" (154–157). Prothero points out that creationists like to argue that complex cells and their organelles such as the bacterial flagellum could not have evolved without special creation, even though it is relatively easy to show the pathways by which complex, eukaryotic cells and their organelles arose. He further states that this "doesn't require anything more complex than living together in peace and harmony" (154). Thus, the best examples that creationists come up with to refute evolution turn out to be easily understood in evolutionary terms if we look to cooperative relationships between organisms, which is the essence of Indigenous thought.

Similar but less obligately mutually dependent situations can be found among animals in corals, anemones, and flatworms that contain endosymbiotic algae, and in lichens, which are fungi with endosymbiotic algae. Plants are even more distinctive; they contain not only mitochondria but also chloroplasts, which are endosymbiotic photosynthetic bacteria that allow them not only to process but also to produce their own nutrition (Margulis 1998). After the origin of life itself, these endosymbiotic mergers are the most significant creative event in the history of life, yet it required no human-style mind or cosmic engineer for it to happen. *Endosymbiosis* means cooperation within, and all that was required was for two different types of organisms to cooperate so that each could exist in a new form, which eventually led to all the life forms recognized by Indigenous peoples.

This most essential relationship for all of life would never have been observed by Indigenous people, because they do not look inside of cells, or even at cells themselves. Discovery of this relationship required the use of powerful microscopes and DNA sequencing before it became well established. Not surprisingly, the concept of endosymbiosis was roundly rejected by Western science when it was first suggested in 1905 by K. S. Merezhkovsky (Margulis 1998). The Western scientific tradition of the time considered bacteria to be "agents of disease, not originators of evolutionary novelty," which is an idea that continues to this day in most people, regardless of culture (Margulis 1998: 53).

Endosymbiosis is a perfect example of a phenomenon that could be predicted and readily understood by Indigenous people, yet it required a highly technologically driven Western scientific set of studies to document its reality. If Indigenous concepts had been merged with those from Western science all along, the endosymbiotic concept would probably have been readily accepted when it was first proposed, because it emerges from recognition and understanding of relationships in a way that flies in the face of the typologically oriented Western scientific tradition, which until recently could not accept that different life forms could combine to form new types of life. Endosymbiosis is a concept based on relationships and co-evolution, which are the basis of much of Indigenous philosophy, belief, and TEK (Barsh 2000; Chapters 2–5), yet because it occurs between tiny cells unobservable without microscopes it would have remained unobserved by Indigenous people because they operate at the level of relationships between organisms. This is a perfect metaphor on which to close this book, a phenomena that requires input from both Indigenous and Western traditions to be recognized, accepted, and fully understood, yet that has not been part of either tradition until very recently.

References

Adams, H. 1918. *Education of Henry Adams: An Autobiography*, Houghton Mifflin, New York, NY.

Alberts, L. 1988. Novel Traces Shattering of Indian Traditions, *Albuquerque Journal*, 23 October 1988, G–8.

Alcoze, T. 2001. Review of *The Ecological Indian. Ethics, Place, and the Environment 4*: 261–266.

Allen, P. G. 1986. *The Sacred Hoop: Recovering the Feminine in American Indian Traditions*, Beacon Press, Boston, MA.

Alessa (Na'ia), L. 2009. What Is Truth? Where Western Science and Traditional Knowledge Converge, in *The Alaska Native Reader: History, Culture, Politics*, ed. M. (Shaa Tl'aa) Williams, Duke University Press, Durham, NC, pp. 246–251.

Alvard, M. S. 2002. Evolutionary Theory, Conservation and Human Environmental Impact, in *Wilderness and Political Ecology: Aboriginal Influences and the Original State of Nature*, eds. C. E. Kay and R. T. Simmons, University of Utah Press, Salt Lake City, UT, pp. 28–43.

Amoss, P. 1987. The Fish God Gave Us: The First Salmon Ceremony Revived, *Arctic Anthropologist 24*: 56–66.

Anderson, E. N. 1996. *Ecologies of the Heart: Emotion, Belief, and the Environment*, Oxford University Press, New York, NY.

Anderson, E. N. 2005. *The Political Ecology of a Yucatec Maya Community*, University of Arizona Press, Tucson, AZ.

Andrewartha, H. G., and L. C. Birch 1954. *The Distribution and Abundance of Animal Numbers*, University of Chicago Press, Chicago, IL.

Annett, C. A., and R. Pierotti. 1989. Chick Hatching as a Trigger for Dietary Switches in Western Gulls, *Colonial Waterbirds 12*: 4–11.

Annett, C. A., and R. Pierotti. 1999. Longterm Reproductive Output and Recruitment in Western Gulls: Consequences of Alternate Foraging Tactics, *Ecology 80*: 288–297.

Armitage, P. 1992. Religious Ideology Among the Innu of Eastern Quebec and Labrador, *Religiologiques 6*: 64–110.

Asquith, P. J. 1986. Anthropomorphism and the Japanese and Western Traditions in Primatology, in *Primate Ontogeny, Cognition and Social Behaviour*, ed. P. Else and R. Lee, Cambridge University Press, New York, NY, pp. 61–71.

Asquith, P. J. 1996. Japanese Science and Western Hegemonies: Primatology and the Limits Set to Questions, in *Naked Science: Anthropological Inquiry Into Boundaries, Power, and Knowledge*, ed. L. Nader, Routledge, New York, NY, pp. 239–256.

Asquith, P. J. 1997. Why Anthropomorphism is *not* Metaphor: Crossing Concepts and Cultures in Animal Behavior Studies, in *Anthropomorphism, Anecdotes, and Animals*, ed. R. W. Mitchell, N. S. Thompson, and H. L. Miles, State University of New York Press, Albany, NY, pp. 37–49.

Atleo, E. R. 2004. *Tsawalk: A Nuu-chah-nulth Worldview,* University of British Columbia Press, Vancouver, BC.

Awiakta, M. 1993. *Selu: Seeking the Corn Mother's Wisdom*, Fulcrum, Golden, CO.

Ball, P. 2001. *Stories of the Invisible: A Guided Tour of Molecules*, Oxford University Press, New York, NY.

Balter, M. 2001. In Search of the First Europeans, *Science 291*: 1722–1725.

Baron, D. 2004. The Beast in the Garden: A modern Parable of Man and Nature. W.W. Norton, New York, NY.

Barsh, R. L. 1997. Forests, Indigenous Peoples, and Biodiversity, *Global Biodiversity* (Canadian Museum of Nature) 7(2): 20–24.

Barsh, R. L. 2000. Taking Indigenous Science Seriously, in *Biodiversity in Canada: Ecology, Ideas, and Action*, ed. S. A. Bocking, Broadview Press, Toronto, ON, pp. 152–173.

Barsh, R. L. N.d. Nonagricultural Peoples and Captive Wildlife: Implications for Ecology and Evolution. Unpublished MS.

Basso, K. 1996. *Wisdom sits in Places*, University of New Mexico Press, Albuquerque, NM.

Bekoff, M. 2000. *The Smile of a Dolphin: Remarkable Accounts of Animal Emotion*, Discovery Books, London.

Bekoff, M. 2002. *Minding Animals: Awareness, Emotions and Heart*, Oxford University Press, New York, NY.

Bekoff, M. 2006. *Animal Passions and Beastly Virtues: Reflections on Redecorating Nature*, Temple University Press, Philadelphia, PA.

Bekoff, M. 2007. *The Emotional Lives of Animals*, New World Library, Novato, CA.

Bell, G. 1980. Costs of Reproduction, *American Naturalist 116*: 45–76.

Berger, J., J. E. Swenson, and I. L. Persson. 2001. Recolonizing Carnivores and Naive Prey: Conservation Lessons from Quaternary Extinctions, *Science 291*: 1036–1039.

Berkes, F. 1999. *Sacred Ecology: Traditional Ecological Knowledge and Resource Management*, Taylor and Francis, Philadelphia, PA.

Berman, J. 2000. Red Salmon and Red Cedar Bark: Another Look at the 19th Century Kwakwaka'wakw Winter Ceremonial, *BC Studies (Special issue) Ethnographic Eyes 125–126*: 53–98.

Boag, P. T., and P. R. Grant. 1981. Intense Natural Selection in a Population of Darwin's Finches (Geospizinae) in the Galapagos, *Science 214*: 82–85.

Botkin, D. 1990. *Discordant Harmonies: A New Ecology for the Twenty-First Century*, Oxford University Press, New York, NY.

Botkin, D. 1991. A New Balance of Nature. *Wilson Quarterly* (Spring): 61–72.

Bowler, P.J. 1990. *Charles Darwin: The Man and his Influence*, Blackwell, Cambridge, UK.

Boyce, M. S. 1984. Restitution of *r*- and *K*-selection as a Model of Density-Dependent Natural Selection, *Annual Review of Ecology and Systematics* 15: 427–447.

Boyd, R., ed. 1999. *Indians, Fire and the Land in the Pacific Northwest*, Oregon State University Press, Corvallis, OR.

Boyd, R. 2006. The Puzzle of Human Sociality, *Science 314*: 1555–1556.

Bradshaw, W. E., and C. M. Holzapfel. 2006. Evolutionary Response to Rapid Climate Change, *Science* 312: 1477–1478.

Bright, W. 1993. *A Coyote Reader*, University of California Press, Berkeley, CA.

Brightman, R. 1993. *Grateful Prey: Rock Cree Human-Animal Relationships*, University of California Press, Berkeley, CA.

Bringhurst, R. 2008. *Everywhere Being Is Dancing: Twenty Pieces of Thinking*, Counterpoint Press, Berkeley, CA.
Brinkerhoff Jackson, J. 1994. *A Sense of Place, a Sense of Time*, Yale University Press, New Haven, CT.
Brody, H. 1982. *Maps and Dreams*, Pantheon Books, New York, NY.
Brody, H. 2000. *The Other Side of Eden: Hunters, Farmers and the Shaping of the World*, North Point Press, New York, NY.
Brooks, D. R., and E. O. Wiley. 1986. *Evolution as Entropy: Toward a Unified Theory of Biology*, University of Chicago Press, Chicago, IL.
Broughton, J. M. 2002. Pre-Columbian Human Impact on California Vertebrates: Evidence from Old Bones and Implications for Wilderness Policy, in *Wilderness and Political Ecology: Aboriginal Influences and the Original State of Nature*, eds. C. E. Kay and R. T. Simmons, University of Utah Press, Salt Lake City, UT, pp. 44–71.
Brown, J. E. 1992. *Animals of the Soul: Sacred Animals of the Oglala Sioux*, Element, Rockport, MA.
Browne, J. 2002. *Charles Darwin: The Origin and After—The Years of Fame*, Alfred A. Knopf, New York, NY.
Browne, J. 2006. *Darwin's Origin of Species: A Biography*, Atlantic Monthly Press, New York, NY.
Bruchac, J. 2003. *Our Stories Remembered: American Indian History, Culture, and Values Through Storytelling*, Fulcrum Press, Golden, CO.
Budiansky, S. 1998. *If a Lion Could Talk*, Free Press, New York, NY.
Buller, G. 1983. Comanche and Coyote, the Culture Maker, in *Smoothing the Ground*, ed. B. Swann, University of California Press, Berkeley, pp. 245–258.
Bulmer, M. 1994. *Theoretical Evolutionary Ecology*, Sinauer Associates, Sunderland, MA.
Burghardt, G. M. 1997. Amending Tinbergen: A Fifth Aim for Ethology, in *Anthropomorphism, Anecdotes, and Animals*, eds. R. W. Mitchell, N. S. Thompson, and H. L. Miles, State University of New York Press, Albany, NY, pp. 254–276.
Cahill, T. 1993. The Muddy Waters of El Peten, in *Pecked to Death by Ducks*. Vintage Departures, Random House, New York, NY, pp. 164–171.
Cajete, G. 1994. *Look to the Mountain: An Ecology of Indigenous Education*, Kivaki Press, Durango, CO.
Cajete, G. 2000. *Native Science: Natural Laws of Interdependence*, Clear Light, Santa Fe, NM.
Caraco, T. 1981. Risk Sensitivity and Foraging Groups, *Ecology* 62: 627–531.
Carroll, S. B. 2005. *Endless Forms Most Beautiful: The New Science of Evo Devo and the Making of the Animal Kingdom*, W. W. Norton, New York, NY.
Carroll, S. 2006. *The Making of the Fittest: DNA and the Ultimate Forensic Record of Evolution*, W. W. Norton, New York, NY.
Carroll, S. 2009. *Remarkable Creatures: Epic Adventures in the Search for the Origins of Species*, Houghton Mifflin Harcourt, New York, NY.
Carson, R. 1962. *Silent Spring*, Houghton Mifflin, Boston, MA.
Casey, S. 2005. *The Devil's Teeth: A True Story of Obsession and Survival among America's Great White Sharks*, Henry Holt, New York, NY.
Cheesman, K., D. French, I. Cheesman, N. Swails, and J. Thomas. 2007. Is there any Common Curriculum for Undergraduate Biology Majors in the 21st century? *Bioscience* 57: 516–522.
Churchill, W. 1995. *Since Predator Came: Notes From the Struggle for American Indian Liberation*, Aigis Press, Littleton, CO.
Clode, D. 2002. *Killers in Eden*, Allen and Unwin, Sydney, Australia.

Clutton-Brock, J. 1984. Dog, in *Evolution of Domesticated Animals*, ed. I. L. Mason, Longman, New York, NY, pp. 198–211.

Clutton-Brock, J. 1995. Origins of the Dog: Domestic and Early History, in *The Domestic Dog,* ed. J. Serpell, Cambridge University Press, Cambridge, UK, pp. 8–20.

Clutton-Brock, T. H., ed. 1988. *Reproductive Success: Studies of Individual Variation in Contrasting Breeding Systems*, University of Chicago Press, Chicago, IL.

Coates, P. 1998. *Nature: Western Attitudes Since Ancient Times*, University of California Press, Berkeley, CA.

Cockburn, A. 1967. *Infectious Diseases: Their Evolution and Eradication*, Thomas Press, Springfield, IL.

Cohen, J. 2009. Key Malaria Parasite Likely Evolved from Chimpanzee Version, *Science 325*: 663.

Cole, L. C. 1954. The Population Consequences of Life History Phenomena, *Quarterly Review of Biology 25*: 103–127.

Coleman, J. T. 2004. *Vicious: Wolves and Men in America*, Yale University Press, New Haven, CT.

Comanche Language and Cultural Preservation Committee. 2003. *Taa Numu Tekwapuʔha Tuboopu (Our Comanche dictionary)*, Comanche Nation, Lawton, OK.

Cook-Lynn, E. 1996. American Indian Intellectualism and the New Indian Story, *American Indian Quarterly 20* (Winter): 71: 57–76.

Cordero, C. 1995. A Working and Evolving Definition of Culture, *Canadian Journal of Native Education 21* (suppl): 7–13.

Cordova, V. F. 2007. *The Native American Philosophy of V.F. Cordova*, eds.K. D. Moore, K. Peters, T. Jojola, and A. Lacy. University of Arizona Press, Tucson, AZ.

Coren, S. 2008. *The Modern Dog*, Free Press, New York, NY.

Coyne, J. A., and H. A. Orr. 2004. *Speciation*, Sinauer Associates, Sunderland, MA.

Crook, J. H. 1964. The Evolution of Social Organization and Visual Communication in the Weaver Birds (Ploeeinae), *Behaviour 10*: 1–178.

Crosby, A. 1972. *The Columbian Exchange: Biological and Cultural Consequences of 1492*, Greenwood Press, Westport, CT.

Cushing, F. H. 1988. *Zuni Fetishes*, Reprint, Las Vegas: K. C. Publications; first published in *Second Annual Report of the Bureau of Ethnology, 1880–1881.* Government Printing Office, Washington, DC.

Darwin, C. 1859. *The Origin of Species by Means of Natural Selection*, Studio Editions, London.

Darwin, C. 1871. *The Descent of Man and Selection in Relation to Sex. 2nd ed.* Reprint, Hurst Publ. New York, NY.

Darwin, C. 1812. *The Expression of Emotions in Man and Animals*, Reprint 1998, Harper Collins, London.

Darwin, C. 1996. *Charles Darwin's Letters: A Selection*, ed. F. Burkhardt, Cambridge University Press, Cambridge, UK.

Davies, R. 1994. *The Cunning Man*, McClelland and Stewart, Toronto.

De Lacy, T., and B. Lawson. 1997. The Uluru-Kakadu Model: Joint Management of Aboriginal-Owned National Parks in Australia, in *Conservation Through Cultural Survival: Indigenous Peoples and Protected Areas*, ed. S. Stevens, Island Press, Washington, DC, pp. 155–187.

Deloria, V., Jr. 1969. *Custer Died for Your Sins*, Collier Macmillan, London.

Deloria, V. 1990. Knowing and Understanding: Traditional Education in the Modern World, *Winds of Change 5(1)*: 12–18.

Deloria, V. 1992. The Spatial Problem of History, in *God is Red*, North American Press, Golden, CO, pp. 114–134.

Deloria, V. 1995. *Red Earth, White Lies: Native Americans and the Myth of Scientific Fact*, Harper and Row, New York, NY.
Deloria, V. 1999a. *For This Land: Writings on Religion in America*, Routledge Press, New York, NY.
Deloria, V. 1999b. *Spirit and Reason: The Vine Deloria, Jr., Reader*, Fulcrum Press, Golden, CO.
Deloria, V. 2000. *Singing for a Spirit: A Portrait of the Dakota Sioux*, Clear Light, Santa Fe, NM.
Deloria, V. 2002. *Evolution, Creationism and other Modern Myths: A Critical Inquiry*, Fulcrum Press, Golden, CO.
Deloria, V., Jr., and D. Wildcat. 2001. *Power and Place: Indian Education in America*, Fulcrum Press, Golden, CO.
Denetclaw, T. H., and W. F. Denetclaw Jr. 1994b. Hantavirus Pulmonary Syndrome in New England and Europe, *New England Journal of Medicine. 331(8)*: 545–548.
Denetclaw, W. F., Jr., and T. H. Denetclaw. 1994a. Is "South-West US Mystery Disease" Caused by Hantavirus? *Lancet. 343(8888)*: 53–54.
Dennett, D. C. 1995. *Darwin's Dangerous Idea: Evolution and the Meanings of Life*, Simon and Schuster, New York, NY.
Diamond, J. 1997. *Guns, Germs, and Steel: The Fates of Human Societies*, W. W. Norton, New York, NY.
Dorit, R. 2007. Biological Complexity, in *Theoretical Ecology*, eds. R. M. May and A. McLean, 3rd ed., Oxford University Press, New York, NY, pp. 231–249.
Dover, G. 2000. Anti-Dawkins, in *Alas Poor Darwin: Arguments Against Evolutionary Psychology*, eds. H. Rose and S. Rose, Harmony Books, New York, NY, pp. 55–78.
Dowie, M. 2005. Conservation Refugees, *Orion 24(6)*: 16–27.
Dreyfus, P. J. 2008. *Our Better Nature: Environment and the Making of San Francisco*, University of Oklahoma Press, Norman, OK.
Druke, M. A. 1980. The Concept of Personhood in 17th and 18th Century Iroquois Ethnopersonality, in *Studies in Iroquois Culture*, ed. N. Bonvillain, Occasional Publications in Northeastern North America no. 6., Man in the Northeast, George's Mill, NH, pp. 59–69.
Dunlap, T. 1988. *Saving America's Wildlife*, Princeton University Press, Princeton, NJ.
Durkheim, E. 1961. *The Elementary Forms of the Religious Life*, Free Press, Glencoe, IL.
Eastman, C. A. (Ohiyesa). 1904. *Red Hunters and the Animal People*, Harper and Row, New York, NY.
Eastman, C. A. (Ohiyesa). 1971. *Indian Boyhood*, Reprint, Dover, New York, NY; first published, McClure, Phillips, New York, 1902.
Eastman, C. A. (Ohiyesa). 1980. *The Soul of the Indian: An Interpretation*, Reprint, Lincoln: University of Nebraska Press; first published, Houghton and Mifflin, Boston, 1911.
Eastman, C. A. (Ohiyesa). 1991. *Old Indian Days*, Reprint, Lincoln: University of Nebraska Press; first published, McClure, New York, 1907.
Eigenbrod, R. 2005. *Traveling Knowledges: Positioning the Im/migrant Reader of Aboriginal Literatures in Canada*, University of Manitoba Press, Winnipeg, Manitoba.
Eldredge, N. 1985. *Time Frames*, Simon and Schuster, New York, NY.
Eldredge, N., and S. J. Gould. 1972. Punctuated Equilibria: An Alternative to Phyletic Gradualism, in *Models in Paleobiology*, ed. T. J. M. Schopf, Freeman, Cooper, San Francisco, pp. 82–115.
Elias, S. A., S. K. Short, and H. H. Birks. 1997. Late Wisconsin Environments on the Bering Land Bridge, *Paleogeography, Paleoclimatology, Paleoecology 136*: 293–308.

Emery, N., and N. Clayton. 2004. The Mentality of Crows: Convergent Evolution of Intelligence in Corvids and Primates, *Science 306*: 1903.
Emlen, S. T. 1995. An Evolutionary Theory of the Family, *Proceedings of the National Academy of Sciences, 92*:8092–8099.
Endler, J. A. 1986. *Natural Selection in the Wild*, Princeton University Press, Princeton NJ.
Erdrich, L. 1984. *Love Medicine*, Holt, Rinehart and Winston, New York, NY.
Erdrich, L. 1988. *Tracks*, Henry Holt, New York, NY.
Erdrich, L. 1995. *The Blue Jay's Dance*, HarperCollins, New York, NY.
Erdrich, L. 1998. *The Antelope Wife*, HarperCollins, New York, NY.
Erdrich, L. 2001. *The Last Report on the Miracles at Little No Horse*, HarperCollins, New York, NY.
Erdrich, L. 2005. *The Painted Drum*, HarperCollins, New York, NY.
Errington, P. 1934. Vulnerability of Bob-white populations to Predation, *Ecology 15*: 110–127.
Errington, P. 1937. What is the Meaning of Predation, in *Annual Report of the Smithsonian Institution for 1936*, Government Printing Office, Washington, DC, pp. 243–252.
Errington, P. 1967. *Of Predation and Life*, Iowa State University Press, Ames, IA.
Errington, P. 1973. *The Red God's Call*, Iowa State University Press, Ames, IA.
Eswaran, V. 2002. A Diffusion Wave Out of Africa, *Current Anthropologist 41*: 741–764.
Fehr, E. and U. Fischbacher. 2003. The Nature of Altruism, *Nature 425*: 785–791.
Feit, H. A. 2007. Myths of the Ecological Whiteman: Histories, Science, and Rights in North American-Native American Relations, in *Native Americans and the Environment: Perspectives on the Ecological Indian,* eds. M. E. Harkin and D. R. Lewis, University of Nebraska Press, Lincoln, NE, pp. 32–51.
Flores, D. 2007. Wars Over Buffalo: Stories Versus Stories on the Northern Plains, in *Native Americans and the Environment: Perspectives on the Ecological Indian,* eds. M. E. Harkin and D. R. Lewis, University of Nebraska Press, Lincoln, NE, pp. 153–170.
Foote, M. 1997. The Evolution of Morphological Diversity, *Annual Review of Ecology and Systematics 28*: 129–152.
Fouts, R. S. 1973. Acquisition and Testing of Gestural Signs in Four young Chimpanzees, *Science 180*: 978–980.
Fragazy, D. M., and S. Perry. 2003. *The Biology of Traditions: Models and Evidence*, Cambridge University Press, Cambridge, UK.
Franz, M. L. von. 1995. *Creation Myths. Revised edition*, Shambhala Books, Boston.
Gerhart, J. C., and M. W. Kirschner. 1997. *Cells, Embryos, and Evolution: Towards a Cellular and Developmental Understanding of Phenotypic Variation and Evolutionary Adaptability*, Blackwell Press, Boston.
Giesel, J. T. 1976. Reproductive Strategies as Adaptations of Life in Temporally Heterogeneous Environments, *Annual Review of Ecology and Systematics 7*: 57–79.
Gilbert, B. 1989. *God Gave Us This Country: Tekamthi and the First American Civil War*, Anchor Books, New York, NY.
Gill, J. L., J. W. Williams, S. T. Jackson, K.B. Lininger, and G. S. Robinson. 2009. Pleistocene Megafaunal Collapse, Novel Plant Communities, and Enhanced Fire Regimes in North America, *Science 326*: 1100–1103.
Giraldeau, L. A. and T. Caraco. 2000. *Social Foraging Theory*, Princeton University Press, Princeton, NJ.
Goebel, T., M. R. Waters, and D. H. O'Rourke. 2008. The Late Pleistocene Dispersal of Modern Humans in the Americas, *Science 319*: 1497–1502.

Gomez-Pompa, A., and A. Kaus. 1992. Taming the Wilderness Myth, *BioScience* *42*: 271–279.
Good, T. P., J. Ellis, C. A. Annett, and R. Pierotti. 2000. Bounded Hybrid Superiority: Effects of Mate Choice, Habitat Selection, and Diet in an Avian Hybrid Zone, *Evolution 54*: 1774–1783.
Goodall, J. 1986. *The Chimpanzees of Gombe*, Harvard University Press, Cambridge, MA.
Goodall, J. 1990. *Through a Window: My Thirty Years With the Chimpanzees of Gombe*, Houghton Mifflin, Boston, MA.
Goodall, J. and M. Bekoff. 2003. *The Ten Trusts: What We Must Do to Care for the Animals We Love,* Harper/Collins Publishers, San Francisco, CA.
Gottfried, R. 1983. *The Black Death: Natural and Human Disaster in Medieval Europe*, Robert Hale, London.
Gould, S. J. 1977. *Ever Since Darwin*, W. W. Norton, New York.
Gould, S. J. 1987. *An Urchin in the Storm: Essays About Books and Ideas*, W. W. Norton, New York.
Gould, S. J. 1999. *Rocks of Ages*, Ballantine, New York.
Gould, S. J. 2002. *The Structure of Evolutionary Theory*, Belknap Press of Harvard University Press, Cambridge, MA.
Gowaty, P. A., ed. 1996. *Feminism and Evolutionary Biology*, Chapman and Hall Press, London.
Grant, B. R., and P. Grant. 1989a. Natural Selection in a Population of Darwin's Finches, *American Naturalist 133*: 377.
Grant, B. R., and P. Grant. 1989b. *Evolutionary Dynamics of a Natural Population: The Large Cactus Finch of the Galapagos*, University of Chicago Press, Chicago, IL.
Grant, B. R., and P. Grant. 1995. Predicting Microevolutionary Responses to Directional Selection on Heritable Variation, *Evolution 49*: 241–251.
Grayson, D. K. 1991. Late Pleistocene Mammalian Extinctions in North America: Taxonomy, Chronology, and Explanation, *Journal of World Prehistory 5*: 193–232.
Grayson, D. K. 2006. Ice Age Extinctions, *Quarterly Review of Biology 81*: 259–264.
Grayson, D. K., and D. J. Meltzer. 2002. Clovis Hunting and Large Mammal Extinction: A Critical Review of the Evidence, *Journal of World Prehistory 16*: 313–359.
Grayson, D. K., and D. J. Meltzer. 2003. A Requiem for Pleistocene Overkill, *Journal of Archaeological Science 30*: 585–593.
Grinnell, G. B. 1962. *Blackfoot Lodge Tales*, University Nebraska Press, Lincoln, NE.
Guthrie, R. D. 1990. *Frozen Fauna of the Mammoth Steppe: The Story of Blue Babe*, University of Chicago Press, Chicago, IL.
Haak, W., et al. 2005. Ancient DNA From the first European Farmers in 7500-year-old Neolithic Sites, *Science 310*: 1016–1018.
Hallowell, A. I. 1960. Ojibwa Ontology, Behavior, and World View, in *Culture in History: Essays in Honor of Paul Radin*, ed. S. Diamond, Columbia University Press, New York, pp. 19–52.
Hamilton, W. D. 1996. *Narrow Roads of Gene Land: The Collected Papers of W. D. Hamilton. Vol. 1: Evolution of Social Behavior*, W. H. Freeman, Oxford, UK.
Hanski, I. 1999. *Metapopulation Ecology*, Oxford University Press, New York.
Hare, B., and M. Tomasello. 2005. Human-Like Social Skills in Dogs, *Trends in Cognitive Science 9*: 439.
Harjo, J., and G. Bird, eds. 1997. *Reinventing the Enemy's Language: Contemporary Native Women's Writings of North America*, W. W. Norton, New York.

Harkin, M. E. 2007. Swallowing Wealth: Northwest Coast beliefs and Ecological Practices, in *Native Americans and the Environment: Perspectives on the Ecological Indian*, eds. M. E. Harkin and D. R. Lewis, University of Nebraska Press, Lincoln, NE, pp. 211–232.

Harkin, M. E. and D. R. Lewis. 2007. *Native Americans and the Environment: Perspectives on the Ecological Indian*, University of Nebraska Press, Lincoln, NE.

Harner, M. 1977. The Ecological Basis for Aztec Sacrifice, *American Ethnologist 4*: 117–135.

Harney, C. 1995. *The Way it Is: One Water . . . One Air . . . One Mother Earth.* Blue Dolphin, Nevada City, CA.

Harrod, H. I. 2000. *The Animals Came Dancing: Native American Sacred Ecology and Animal Kinship*, University of Arizona Press, Tucson, AZ.

Hausfater, G., and S. B. Hrdy. 1984. *Infanticide: Comparative and Evolutionary Perspectives*, Aldine Press, New York.

Havel, I. M. 1996. Scale Dimensions on Nature, *International Journal of General Systematics 24*: 295–324.

Heinrich, B. 1989. *Ravens in Winter*, Simon and Schuster, New York.

Heinrich, B. 2000. *The Mind of the Raven: Investigations and Adventures with Wolf-birds*, HarperCollins, New York.

Henriksen, G. 2009. *I Dreamed the Animals: Kaniuekutat, The Life of an Innu Hunter*, Berghahn Books, New York.

Herman, L. 1986. Cognition and Language Competencies of Bottle-nosed Dolphins, in *Dolphin Cognition and Behavior: A Comparative Approach*, eds. R. J. Schusterman, J. Thomas, and F. G. Woods, Lawrence Earlbaum, Hillsdale, NJ, pp. 221–252.

Herzog, Werner, director. 2005. *Grizzly Man*. Documentary DVD. 103 min. Lions Gate Films.

Hesse, M. 1974. *The Structure of Scientific Inference*, University of California Press, Berkeley, CA.

Hoffman, A. A., and P. A. Parsons. 1997. *Extreme Environmental Change and Evolution*, Cambridge University Press, New York.

Holgate, P. 1967. Population Survival and Life History Phenomena, *Journal of Theoretical Biology 14*: 1–10.

Holt, R. D. 2006. Asymmetry and Stability, *Nature 442*: 252–253.

Holt, S. J. 1975. The Concept of Maximum Sustainable Yield (MSY) and its Application to Whaling, *FAO/UN Scientific Consultation on Marine Mammals.* Document ACMRR/MM/SC/4.

Honderich, T. 2005. *The Oxford Companion to Philosophy*, 2nd ed. Oxford University Press, New York.

Houston, A. I., and J. M. McNamara. 1999. *Models of Adaptive Behaviour: An Approach Based on State*, Cambridge University Press, Cambridge, MA.

Howard, D. J., and S. H. Berlocher, eds. 1998. *Endless Forms: Species and Speciation*, Oxford University Press, New York.

Howitt, R. 2001. *Rethinking Resource Management: Justice, Sustainability, and Indigenous Peoples,* Routledge Press, New York.

Hughes, J.D. 1996. *North American Indian Ecology*, Texas Western Press, El Paso, TX.

Hunt, G. L., Jr., and M. W. Hunt. 1976. Gull Chick Survival: The Significance of Growth Rates, Timing of Breeding, and Territory Size, *Ecology 57*: 62–75.

Huntington, H. P. 1992. *Wildlife Management and Subsistence Hunting in Alaska.* University of Washington Press, Seattle.

Huntington, H. P., and N. I. Myrmin. 1996. *Traditional Ecological Knowledge of Beluga Whales: An Indigenous Knowledge Pilot Project in the Chukchi and Northern Bering Seas*, Inuit Circumpolar Conference, Anchorage, AK.

Huxley, T. H. 1894. *Evolution and Ethics*, Appleton and Co., New York, NY.
Hyde, G. E. 1968. *Life of George Bent, Written From His Letters*, University of Oklahoma Press, Norman, OK.
Iglauer, E. 1988. *Fishing with John*, Harbour Press, Madeira Park, BC, Canada.
"Iron Eyes Cody" (Espera DeCorti). *Character Actors: The Names You Don't Remember, the Faces You Can't Forget.* http://www.what-a-character.com/.
Isenberg, A. C. 2000. *The Destruction of the Bison: An Environmental History, 1750–1920*, Cambridge University Press, New York.
Itani, J., and A. Nishimura. 1973. The Study of Infrahuman Culture in Japan: A Review, in *Precultural Primate Behavior*, ed. E. W. Menzel, Karger Press, Basel, Switzerland, pp. 26–50.
Jackson, W. 1994. *Becoming Native to this Place*, University of Kentucky Press, Lexington, KY.
Johannes, R. E. 1989. *Traditional Ecological Knowledge: A Collection of Essays*, IUCN (World Conservation Union), Gland, Switzerland.
Johnson, C. 2009. Megafaunal Decline and Fall, *Science 326*: 1072–1073.
Jones, D. 2005. Personal Effects, *Nature 438*: 14–16.
Kawagley, A. O. 1995. *A Yupiaq Worldview: A Pathway to Ecology and Spirit*, Waveland Press, Prospect Heights, IL.
Kawaharada, D., ed. 1994. *Nanaue, the Shark Man andOther Hawaiian Stories*, Kalamaku' Press, Honolulu, HI.
Kay, C. E. 2002a. Are Ecosystems Structured From the Top Down or Bottom Up? A New Look at an Old Debate, in *Wilderness and Political Ecology: Aboriginal Influences and the Original State of Nature*, eds. C. E. Kay and R. T. Simmons, University of Utah Press, Salt Lake City, UT, pp. 215–237.
Kay, C. E. 2002b. Afterword: False Gods, Ecological Myths and Biological Reality, in *Wilderness and Political Ecology: Aboriginal Influences and the Original State of Nature*, eds. C. E. Kay and R. T. Simmons, University of Utah Press, Salt Lake City, UT, pp. 238–262.
Kay, C. E. and R. T. Simmons, eds. 2002. *Wilderness and Political Ecology: Aboriginal Influences and the Original State of Nature*, University of Utah Press, Salt Lake City, UT.
Kennedy, J. S. 1992. *The New Anthropomorphism*, Cambridge University Press, Cambridge, UK.
Kidwell, C. S., H. Noley, and G. E. Tinker. 2001. *A Native American Theology*, Orbis Books, Maryknoll, NY.
Kidwell, C. S., and A. Velie. 2005. *Native American Studies*, University of Nebraska Press, Lincoln, NE.
Kingsland, S. E. 1985. *Modeling Nature: Episodes in the History of Population Ecology*, University of Chicago Press, Chicago, IL
Kingsland, S. E. 1991. Defining Ecology as a Science, in *Foundations of Ecology: Classic Papers with Commentaries*, eds. L. A. Real and J. H. Brown, University of Chicago Press, Chicago, IL, pp. 1–13.
Kingsland, S. E. 2005. *The Evolution of American Ecology, 1890–2000*, Johns Hopkins University Press, Baltimore, MD.
Kirschner, M. W., and J. C. Gerhart. 2005. *The Plausibility of Life: Resolving Darwin's Dilemma*, Yale University Press, New Haven, CT.
Klubnikin, K., C. Annett, M. Cherkasova, M. Shishin, and I. Fotieva. 2000. The Sacred and the Scientific: Traditional Ecological Knowledge in Siberian River Conservation, *Ecological Applications 10*: 1296–1306.
Krebs, J. R., and N. B. Davies. 1984. *Behavioral Ecology: An Evolutionary Approach*, Blackwell Scientific Press, New York, NY.
Krech, S., III. 1999a. *The Ecological Indian: Myth and History*, W. W. Norton, New York, NY.

Krech, S. 1999b. Playing with Fire, *New Scientist*, 23 October: 5656.

Krech, S., III, and C. Merchant, eds. 2004. *Encyclopedia of World Environmental History*, Berkshire Press, New York, NY.

Kruuk, H. 2002. *Hunter and Hunted: Relationships between Carnivores and People*, Cambridge University Press, Cambridge.

Kurlansky, M. 1997. *Cod: A Biography of the Fish that Changed the World*, Walker Press, New York, NY.

Lack, D. 1954. *The Natural Regulation of Animal Numbers*, Oxford University Press, New York, NY.

Lack, D. 1968. *Ecological Adaptations for Breeding in Birds*, Chapman and Hall, London.

LaFlesche, F. 1995. *The Osage and the Invisible World. Civilization of the American Indian vol. 21*, University of Oklahoma Press, Norman, OK.

Laland, M. and B. G. Galef. 2009. *The Question of Animal Culture*, Harvard University Press, Cambridge, MA.

LaLonde, C. 2002. *Grave Concerns, Trickster Turns: The Novels of Louis Owens*, University of Oklahoma Press, Norman, OK.

Langdon, S. J. 2007. Sustaining a Relationship: Inquiry Into the Emergence of a Logic of Engagement with Salmon among the Southern Tlingit, in *Wilderness and Political Ecology: Aboriginal Influences and the Original State of Nature*, eds. C. E. Kay and R. T. Simmons, University of Utah Press, Salt Lake City, UT, pp. 233–273.

Lapinski, M. 2005. *Death in the Grizzly Maze: The Timothy Treadwell Story*, Falcon Books, Helena, MT.

Larkin, P.A. (1977) An Epitaph for the Concept of MSY, *Transactions of the American Fisheries Society 106*: 1–11.

Lawton, M. 1996. The Mask of Theory and the Face of Nature, in *Feminism and Evolutionary Biology*, ed. P. A. Gowaty, Chapman and Hall, London, pp. 261–275.

Le Guin, U. K. 2004. *The Wave in the Mind: Talks and Essays on the Writer, the Reader, and the Imagination*, Shambala Books, Boston, MA.

Lehr, J., ed. 1992. *Rational Readings on Environmental Issues*, Van Nostrand Reinhold, New York, NY.

Lemman, N. 1999. Buffaloed, *New Yorker, 13 September 1999*: 98–101.

Lemonick, M. D., and A. Dorfman. 2006. Who Were the First Americans? *Time, 13 March 2006*, 44–52.

Leopold, A. 1948. *A Sand County Almanac*, Cambridge University Press, Cambridge.

Levins, R. 1968: *Evolution in Changing Environments: Some Theoretical Explorations*, Princeton University Press, Princeton, NJ.

Levy, S. 2005. Rekindling Native Fires, *Bioscience 55*: 303–308.

Lewontin, R. 1997. Billions and Billions of Demons: A Review of *The Demon-Haunted World: Science as a Candle in the Dark* by Carl Sagan. *NY Times Book Reviews 1/9/97*.

Lewontin, R. 2001. *The Triple Helix: Gene, Organism, and Environment*, Harvard University Press, Cambridge, MA.

Lewis, H. T. 1989. A Parable of Fire: Hunter Gatherers in Canada and Australia, in *Traditional Ecological Knowledge: A Collection of Essays*, ed. R. E. Johannes, IUCN, Cambridge, pp. 7–20.

Lewis, M. 1994. *Green Delusions*, University of North Carolina Press, Raleigh.

Logsdon, G. 1972. John Joseph Matthews—A Conversation, *Nimrod 16.2*: 70–75.

Lopez, B. 1990. *Crow and Weasel*, North Point Press, San Francisco.

Longworth, P. 1997. *The Making of Eastern Europe: From Prehistory to Postcommunism, 2nd ed.* Macmillan, London.
Lyons, O. 1984. Our Mother Earth, *Parabola 7(1)*: 91–93.
MacArthur, R. H. 1972. *Geographical Ecology*, Harper and Row, New York.
Mailhot, J. 1994. *Traditional Ecological Knowledge: The Diversity of Knowledge Systems and their Study*, Great Whale Environmental Assessment 4. Great Whale Review, Montreal.
Mander, J. 1991. *The Absence of the Sacred*, Sierra Club Books, San Francisco, CA.
Mangel, M. and C. W. Clark. 1988. *Dynamic Modeling in Behavioral Ecology*, Princeton University Press, Princeton, NJ.
Mann, C. C. 2002. 1491. *Atlantic Monthly, March*, 41–53.
Mann, C. C. 2005. *1491: New Revelations of America Before Columbus*, Knopf, New York.
Manning, R. 2004. *Against the Grain: How Agriculture has Hijacked Civilization*, North Point Press, New York.
Margulis, L. 1998. *Symbiotic Planet: A New View of Evolution*, Basic Books, New York.
Marsh, G. P. 1864. *Man and Nature: Or, Physical Geography as Modified by Human Action*, Charles Scribner, NY.
Marshall, E. 2001a. Pre-Clovis Sites Fight for Acceptance, *Science 291*: 1730–1732.
Marshall, E . 2001b. Clovis First, *Science 291*: 1731.
Marshall, J., III. 1995. *On Behalf of the Wolf and the First Peoples*, Red Crane Books, Santa Fe, NM.
Marshall, J., III. 2001. *The Lakota Way*, Viking Compass Books, New York.
Marshall, J., III. 2005. *Walking With Grandfather: Teachings From Lakota Wisdom Keepers,* Sounds True Press, Louisville, CO.
Marshall-Thomas, E. 1994. *The Tribe of Tiger: Cats and Their Culture*, Simon and Schuster, New York, NY.
Martin, C. L. 1978. *Keepers of the Game: Indian-Animal Relationships and the Fur Trade*, University of California Press, Berkeley.
Martin, C. L. 1999. *The Way of the Human Being*, Yale University Press, New Haven, CT.
Martin, J. W. 2000. *The Land Looks after Us: A History of Native American Religion*, Oxford University Press, New York, NY.
Martin, P. S. 2005. *Twilight of the Mammoths: Ice-Age Extinctions and the Rewilding of America*, University of California Press, Berkeley.
Martinez, D. 1994. Traditional Environmental Knowledge connects Land and Culture, *Winds of Change 9(4)*: 89–94.
Marzluff, J. and T. Angell. 2005. *In the Company of Crows and Ravens*. Yale University Press, New Haven, CT.
May, R. M. 1973. *Stability and Complexity in Model Ecosystems*, Princeton University Press, Princeton, NJ.
May, R. M., and A. McLean. 2007. *Theoretical Ecology, 3rd ed*, Oxford University Press, New York.
Mayr, E. 1961. *Animal Evolution*, Harvard University Press, Cambridge, MA.
Mayr, E. 1997. *This is Biology: The Science of the Living World*, Belknap Press, Cambridge, MA.
Mayr, E. and W. B. Provine. 1998. *The Evolutionary Synthesis: Perspectives on the Unification of Biology*, Harvard University Press, Cambridge, MA.
McClellan, C. 1975. *My Old People Say: An Ethnographic Survey of the Southern Yukon*, Ottawa, National Museum of Canada.

McClintock, W. 1910. *The Old North Trail; or, Life, Legends, and Religion of the Blackfeet Indians*, Macmillan Press, London.
McComb, K., C. Moss, S. M. Durant, L. Baker, and S. Saylelel. 2001. Matriarchs as Repositories of Social Knowledge in African Elephants, *Science 292*: 491–494.
McGaa, E. 1995. *Native Wisdom: Perceptions of the Natural Way*, Four Directions Pub, Minneapolis, MN.
McIntosh, James. 1974. *Thoreau as Romantic Naturalist: His Shifting Stance Towards Nature*, Cornell University Press, Ithaca, NY.
McIntyre, R. 1995. *War Against the Wolf: America's Campaign to Exterminate the Wolf*, Voyageur Press, Stillwater, MN.
McKervill, H. W. 1992. *The Salmon People, 25th Anniversary Edition*, Whitecap Books, Vancouver, BC.
McLuhan, T. C. 1971. *Touch the Earth: A Self-Portrait of Indian Existence*, Promontory Press, New York.
Mech, L. D. 1998. Who's Afraid of the Big Bad Wolf?—Revisited, *International Wolf 8(1)*: 8–11.
Meggitt, M. J. 1965. The Association between Australian Aborigines and Dingoes, in *Man, Culture and Animals: The Role of Animals in Human Ecological Adjustments*, eds. A. Leeds and A. P. Vayda, American Association for the Advancement of Science, Washington, DC, pp. 7–26.
Menzies, C. R., ed. 2006. *Traditional Ecological Knowledge and Natural Resource Management*, University of Nebraska Press, Lincoln, NE.
Mertz, D. B. 1971a. Life History Phenomena in Increasing and Decreasing Populations, in *Statistical ecology, vol. 2*, eds. G. P. Patil, E. C. Pielou, and W. E. Waters, Pennsylvania State University Press, University Park, PA, pp. 361–399.
Mertz, D. B. 1971b. The Mathematical Demography of the California Condor Population, *American Naturalist 105*: 437–453.
Mesoudi, A., A. Whiten, and K. Laland. 2004. Is Human Cultural Evolution Darwinian? Evidence Reviewed From the Perspective of *The Origin of Species, Evolution 58*: 1–11.
Miller, J. D., E. C. Scott, and S. Okamoto. 2006. Public Acceptance of Evolution, *Science 313*: 765–766.
Miller, W. R., and R. M. Brigham. 1988. "Ceremonial" Gathering of Black-billed Magpies, *Pica pica*, after the Sudden Death of a Conspecific, *Murrelet 69*: 78–79.
Minta, S. C., K. A. Minta, and D. F. Lott. 1992. Hunting Associations Between Badgers and Coyotes, *Journal of Mammalogy 73*: 814–820.
Mitchell, R. W., N. S. Thompson, and H. L. Myles, eds. 1997. *Anthropomorphism, Anecdotes and Animals*, State University of New York Press, Albany, NY.
Momaday, N. S. 1968. *House Made of Dawn*, Harper and Row, New York.
Momaday, N. S. 1969. *The Way to Rainy Mountain*, University of New Mexico Press, Albuquerque, NM.
Momaday, N. S. 1989. *The Ancient Child*, Doubleday, New York.
Monod, J. 1979. *Chance and Necessity*, Collins, London.
Morey, D. 1986. Studies on Amerindian Dogs: Taxonomic Analysis of Canid Crania from the Northern Plains, *Journal of Archaeological Science 13*: 119–145.
Morey, D. 1994. The Early Evolution of the Domestic Dog, *American Scientist 82*: 336–347.
Mourning Dove. 1990. *Coyote Stories*, University of Nebraska Press, Lincoln, NE.
Mowat, F. 1997. Farley Mowat, Harold Horwood: Conversations, in *Echoing Silence: Essays on Arctic Narrative*, ed. J. Moss, University of Ottawa Press, Ottawa, pp. 107–112.

Mt. Pleasant, J. 2001. The Three Sisters Care for the Land and the People, in *Science and Native American Communities: Legacies of Pain, Visions of Promise*, ed. K. James, University of Nebraska Press, Lincoln, NE, pp. .126–134

Mt. Pleasant, J. 2006. The Science Behind the Three Sisters Mound System: An Agronomic Assessment of an Indigenous Agricultural System in the Northeast, Ch. 38 in *Histories of Maize: Multidisciplinary Approaches to the Prehistory, Linguistics, Biogeography, Domestication, and Evolution of Maize*, eds. J. E. Staller, R. H. Tykot, and B. F. Benz, Elsevier Academic Press, New York, NY.

Mulvaney, J., and J. Kamminga. 1999. *Prehistory of Australia*, Allen and Unwin, Sydney, Australia.

Murphy, G. I. 1968. Pattern in Life History and the Environment, *American Naturalist 102*: 390–404.

Myers, R. A., and B. Worm. 2003. Rapid Worldwide Depletion of Predatory Fish Communities, *Nature 423*: 290–283.

Myers, R. A., and B. Worm. 2005. *Proceedings of the Royal Society of London; Series B: Biological Science 360:* 13–20.

Nadasdy, P. 2003. *Hunters and Bureaucrat: Power, Knowledge, and Aboriginal State Relations in the Southwest Yukon*, UBC Press, Vancouver.

National Geographic. 1995. *Yellowstone: Realm of the Coyote.* Written by Rob Goldberg and John Rubin. VHS video. 60 min. National Geographic Television and Film.

National Geographic. 2006. *Thunderbeast.* Produced by Bob Landis and Kathy Pasternak. DVD documentary. 52 min. National Geographic Television and Film. Shown on National Geographic Channel June 16, 2006.

Nee, S. 2007. Metapopulations and Their Spatial Dynamics, in *Theoretical Ecology*, ed. R. M. May and A. McLean, 3rd ed., Oxford University Press, New York, pp. 35–45.

Nelson, R. 1983. *Make Prayers to the Raven*, University of Chicago Press, Chicago, IL.

Nerburn, K. 1994. *Neither Wolf Nor Dog: On Forgotten Roads With an Indian Elder*, New World Library, Novato, CA.

Nerburn, K. 2009. *The Wolf at Twilight: An Indian Elder's Journey Through a Land of Ghosts and Shadows*, New World Library, Novato, CA.

Nesper, L. 2002. *The Walleye War: The Struggle for Ojibwe Spearfishing and Treaty Rights*, University of Nebraska Press, Lincoln, NE.

Nettleship, D. N. 1972. Breeding Success of the Common Puffin (*Fratercula arctica* L.) on Different Habitats at Great Island, Newfoundland, *Ecological Monographs 42*: 239–268.

Newcomb, F. J. 1990. *Navajo Folk Tales*, Reprint, University of New Mexico Press; first published by Museum of Navajo Ceremonial Arts, 1967.

Newton, I. 1989. *Lifetime Reproduction in Birds*, Academic Press, New York.

Nicholson, M. 1959. *Mountain Gloom and Mountain Glory*, Cornell University Press, Ithaca, NY.

Noriega, J. 1992. American Indian Education in the United States: Indoctrination for Subordination to Colonialism, in *The State of Native America*, ed. M. A. Jaimes, South End Press, Boston, MA, pp. 371–402.

O'Neill, D. 2004. *The Last Giant of Beringia: The Mystery of the Bering Land Bridge*, Westview Press, Boulder, CO.

Ohnuki-Tierney, E. 1995. Representations of the Monkey in Japanese Culture, in *Ape, Man, Apeman: Changing Views Since 1600*, eds. R. Corbey and B. Theunissen, Leiden, the Netherlands, pp. 297–308.

Owens, L. 1991. *Wolfsong*, West End Press, Albuquerque, NM.

Owens, L. 1992a. *Other Destinies: Understanding the American Indian Novel*, University of Oklahoma Press, Norman, OK.

Owens, L. 1992b. *The Sharpest Sight*, University of Oklahoma Press, Norman, OK.

Owens, L. 1994. *Bone Game*, University of Oklahoma Press, Norman, OK.

Owens, L. 1996. *Nightland,* Penguin Books, New York.

Owens, L. 1998. *Mixedblood Messages: Literature, Film Family, Place*, University of Oklahoma Press, Norman, OK.

Owens, L. 1999. *Dark River*, University of Oklahoma Press, Norman, OK.

Owens, L. 2001. *I Hear the Train: Reflections, Inventions, Refractions*, University of Oklahoma Press, Norman, OK.

Packer, C. 1988. Constraints on the Evolution of Reciprocity: Lessons from Cooperative Hunting, *Ethology and Sociobiology* 9:137–149.

Packer, C. 2009. Rational Fear, *Natural History 118(4)*:42–47.

Packer, C., and L. Ruttan. 1988. The Evolution of Cooperative Hunting, *American Naturalist 132*: 159–194.

Papanikolas, Z. 1995. *Trickster in the Land of Dreams*, University of Nebraska Press, Lincoln, NE.

Parfit, M. 2000. Hunt for the First Americans, *National Geographic 198(6)*: 40–67.

Patterson, F. G. P. 1980. Innovative Uses of Language by a Gorilla: A Case Study, in *Children's Language*, ed. K. E. Nelson, vol. 2, Gardner Press, New York, pp. 497–561.

Patterson, F. G. P., and E. Linden. 1981. *The Education of Koko*, Holt, Rinehart and Winston, New York.

Pavlik, S. 1997. American Indian Spirituality, Traditional Knowledge, and the "Demon-haunted" World of Western Science, *American Indian Culture and Research Journal 21*: 281–293.

Payne, K. 2000. *Silent Thunder: In the Presence of Elephants*, Viking Press, New York.

Payne, K. 2003. Sources of Social Complexity in the Three Elephant Species, in *Animal Social Complexity: Intelligence, Culture and Individualized Societies*, eds. F. B. de Waal and P. L. Tyack, Harvard University Press, Cambridge, MA, pp. 57–85 .

Pearce, F. 2007. *With Speed and Violence: Why Scientists Fear Tipping Points in Climate Change*, Beacon Press, Boston.

Penn, W. S. 1995. *All My Sins are Relatives*, University of Nebraska Press, Lincoln, NE.

Penn Hilden, P. 1995. *When Nickels Were Indians: An Urban Mixed Blood Story*, Smithsonian Institution Press, Washington, DC.

Pennisi, E. 2006. Social Animals prove their Smarts, *Science 312*: 1734–1738.

Pepperburg, I. M. 1981. Functional Vocalizations by an African Grey Parrot, *Zeitschrift fur Tierpsychologie 55*: 139–160.

Pepperburg, I. M. 1983. Cognition of the African Grey Parrot: Preliminary Evidence for Auditory/Vocal Comprehension of the Class Concept, *Animal Learning and Behavior 11*: 179–185.

Peterson, D. 2008. *Jane Goodall: The Woman who Redefined Science*, Mariner Books, Houghton Mifflin, Boston, MA.

Petto, A. J., and L. R. Godfrey, eds. 2007. *Scientists Confront Intelligent Design and Creationism*, W. W. Norton, New York.

Pianka, E. R. 1970. On "r" and "K" Selection, *American Naturalist 104*: 592–597.

Pierotti, R. 1981. Male and Female Parental Roles in the Western Gull Under Different Environmental Conditions, *Auk 98*: 532–549.

Pierotti, R. 1982. Habitat Selection and its Effect on Reproductive Output in the Herring Gull in Newfoundland, *Ecology 63*: 854–868.

Pierotti, R. 1983. Gull-puffin Interactions, *Biological Conservation* 26: 1–14.
Pierotti, R. 1987. Behavioral Consequences of Habitat Selection in the Herring Gull, *Studies in Avian Biology 10*: 119–128.
Pierotti, R. 1988a. Associations Between Marine Birds and Marine Mammals in the Northwest Atlantic, in *Seabirds and other Marine Vertebrates: Commensalism, Competition, and Predation*, ed. J. Burger, Columbia University Press, New York, pp. 31–58.
Pierotti, R. 1988b. Interactions Between Gulls and Otariid Pinnipeds: Competition, Commensalism, and Cooperation, in *Seabirds and Other Marine Vertebrates: Commensalism, Competition, and Predation*, ed. J. Burger, Columbia University Press, New York, pp. 213–239.
Pierotti, R. 1989. Intergenerational Conflicts in Species of Birds with Precocial Offspring, *Proceedings of the International Ornithological Congress 19*: 1024–1029.
Pierotti, R. 1991a. Individual Variation and the Costs of Reproduction, *Condor 93*: 1039–1040.
Pierotti, R. 1991b. Adoption vs. Infanticide: An Intergenerational Conflict in Birds and Mammals, *American Naturalist 138*: 1140–1158.
Pierotti, R. 2004. Animal Diseases as Environmental Factors, in *Encyclopedia of World Environmental History*, ed. S. Krech and C. Merchant, Berkshire Press, New York, pp. 313–316.
Pierotti, R. 2005a. Hunting, Religious Restrictions and Implications, in *Encyclopedia of Native American Religious Practices*, ABC CLIO Press, Broomfield, CO, pp. 391–400.
Pierotti, R. 2005b. Religious Overtones in the Writings of Louis Owens, in *Encyclopedia of Native American Religious Practices*, ABC CLIO Press, Broomfield, CO, pp. 723–730.
Pierotti, R., and C. A. Annett. 1987. Reproductive Consequences of Specialization and Switching in an Ecological Generalist, in *Foraging Behavior*, eds. A. C. Kamil, J. R. Krebs, and H. R. Pulliam, Plenum Press, New York, pp. 417–442.
Pierotti, R., and C. A. Annett. 1990. Diet and Reproductive Performance in Seabirds, *Bioscience 40*: 568–574.
Pierotti, R., and C. A. Annett. 1991. Diet Choice in the Herring Gull: Effects of Constraints imposed by Reproduction and Ecology, *Ecology* 72: 319–328.
Pierotti, R., and C. A. Annett. 1993. Hybridization and Male Parental Care in Birds, *Condor 95*: 670–679.
Pierotti, R., and C. A. Annett. 1994. Patterns of Aggression in Gulls: Asymmetries and Tactics in Different Roles, *Condor 96*: 590–599.
Pierotti, R., C. A. Annett, and J. L. Hand. 1996. Male and Female Perceptions of Pair-bond Dynamics: Monogamy in the Western Gull, in *Feminism and Evolutionary Biology*, ed. P. A. Gowaty, Chapman and Hall Press, London, 261–275.
Pierotti, R., and E. C. Murphy. 1987. Intergenerational Conflicts in Gulls, *Animal Behaviour 35*: 435–444.
Pierotti, R., and D. Wildcat. 1997a. Evolution, Creation, and Native Traditions, *Winds of Change 12(2)*: 73–77.
Pierotti, R., and D. Wildcat. 1997b. The Science of Ecology and Native American Traditions, *Winds of Change 12(4)*: 94–98.
Pierotti, R., and D. Wildcat. 1999a. Connectedness of Predators and Prey: Native Americans and Fisheries Management, *Fisheries 24(4)*: 22–23.
Pierotti, R., and D. Wildcat. 1999b. Traditional Knowledge, Culturally Based Worldviews and Western Science, in *Cultural and Spiritual Values of Biodiversity*, ed. D. Posey, U. N. Environment Program Intermediate Technology Publications, London, pp. 192–199.

Pierotti, R., and D. Wildcat. 2000. Traditional Ecological Knowledge: The Third Alternative, *Ecological Applications 10*: 1333–1340.

Pierotti, R., and D. Wildcat. 2001. Being Native to this Place, in *American Indians in History, 1870–2000*, ed. S. Evans, Greenwood, Westport, CT, pp. 3–16.

Pierotti, R., D. Wildcat, V. Deloria, and R. Barsh. N.d. Unpublished letter to Science, 24 July 2003, commenting on Berger, Swenson, and Persson 2001.

Power, S. 1994. *The Grass Dancer*, G. P. Putnam, New York.

Preece, R. 1999. *Cultural Myths, Cultural Realities*, University of British Columbia Press, Vancouver.

Prothero, D. R. 2007. *Evolution: What the Fossils Say and Why it Matters*, Columbia University Press, New York.

Raban, J. 1999. *Passage to Juneau: The Sea and its Meanings*, Pantheon Books, New York.

Ramsey, J. 1977. *Coyote Was Going There: Indian Literature of the Oregon Country*, University of Washington Press, Seattle, WA.

Ranco, D. 2007. The Ecological Indian and the Politics of Representation: Critiquing the Ecological Indian in the Age of Ecocide, in *Native Americans and the Environment: Perspectives on the Ecological Indian*, eds, M. E. Harkin and D. R. Lewis, University of Nebraska Press, Lincoln, NE, pp. 32–51.

Rappaport, R. A. 1971. The Sacred in Human Evolution, *Annual Review of Ecology and Systematics* 2: 23–44.

Rasmussen, K. 1929. The Intellectual Culture of the Iglulik Eskimos, *Vol. 7 of Report of the Fifth Thule Expedition, 1921–1924*. Gyldendalske Boghandel, Copenhagen.

Redman, C. 1999. *Human Impacts on Ancient Environments*, University of Arizona Press, Tucson, AZ.

Reichel-Dolmatoff, G. 1996. *The Forest Within: The Worldview of the Tukano Amazonian Indians*, Council Oak Books, Tulsa, OK.

Reznick, D., F. Shaw, F. H. Rodd, and R. G. Shaw. 1997. Evaluation of the Rate of Evolution in Natural Populations of Guppies, *Science 275*: 1934.

Richardson, S. 1999. Review of *The Ecological Indian, Discovery 9/99*: 98.

Ricklefs, R. 1998. *The Economy of Nature*, W. H. Freeman, New York.

Ridley, M. 2004. *Evolution, 3rd ed.*, Blackwell, Malden, MA.

Riedman, M. 1988. Adoption and Alloparental Care in Mammals and Birds, *Quarterly Review of Biology 57*: 405–435.

Rockwell, D. 1991. *Giving Voice to Bear: North American Indian Rituals, Myths, and Images of the Bear*, Roberts-Rinehart, Niwott, CO.

Roeder, A. S., and D. Wilson. 1995. *Mammalian Species of the World. 2nd ed.*, Smithsonian Institute Press, Washington, DC.

Rohman, C. 2009. *Stalking the Subject: Modernism and the Animal.* Columbia University Press, New York, NY.

Roff, D. 1992. *The Evolution of Life Histories: Theory and Analysis*, Chapman and Hall, London.

Rooney, N., K. McGann, G. Gellner, and J. C. Moore. 2006. Structural Asymmetry and the Stability of Diverse Food Webs, *Nature 442*: 265–269.

Rose, W. 1992. The Great Pretenders: Further Reflections on Whiteshamanism, in *The State of Native America*, ed. M. A. Jaimes, South End Press, Boston, pp. 403–422.

Ross, R. 1992. *Dancing with a Ghost: Exploring Indian Reality*, Reed Books, Canada.

Rothstein, S. I. and R. Pierotti. 1989. Definitions and the Genetic Basis of Beneficent Behavior, *Ethology and Sociobiology 10*: 453–456.

Ruse, M. 2008. *Charles Darwin*, Blackwell Publishing, Malden, MA.

Sablin, M. V., and G. A. Khlopachev. 2002. The Earliest Ice-Age Dogs: Evidence from Eliseevichi Island, *Current Anthropology 43*:795–799.

Sagan, C. 1995. *The Demon-Haunted World: Science as a Candle in the Dark*, Random House, New York.

Sale, K. 1991. *The Conquest of Paradise: Christopher Columbus and the Columbian Legacy,* Alfred A. Knopf, New York.

Salmon, E. 2000. Kincentric Ecology: Indigenous Perceptions of the Human-Nature Relationship, *Ecological Applications 10*: 1327.

Schleidt, W. M. 1998. Is Humaneness Canine? *Human Ethology Bulletin 13(4)*: 3–4.

Schlesier, K. H. 1987. *The Wolves of Heaven: Cheyenne Shamanism, Ceremonies, and Prehistoric Origins*, Civilization of the American Indian series, no. 183. University of Oklahoma Press, Norman, OK.

Schubnell, M. 1985. *N. Scott Momaday: The Cultural and Literary Background*, University of Oklahoma Press, Norman, OK.

Schweitzer, A. 1933. *Out of My Life and Thought: The Autobiography of Albert Schweitzer*, Trans. Antje Bultmann Lemke. Henry Holt, New York.

Schweitzer, A. 1950. *The Animal World of Albert Schweitzer: Jungle Insights into Reverence for Life*, Ecco Press, Hopewell, NJ.

Scott, E. C. 1997. Antievolution and Creationism in the United States, *Annual Review of Anthropology* 26: 263–289.

Scott, E. C. 2007. Creation Science Lite: "Intelligent Design" as the New Anti-evolutionism, in *Scientists Confront Intelligent Design and Creationism*, eds. A. J. Petto and L. R. Godfrey, W. W. Norton, New York, 59–109.

Scott, E. C., and K. Padian. 1997. The new Antievolutionism—What to Do About It, *Trends in Ecology and Evolution* 12: 84.

Serpell, J. 1995. From Paragon to Pariah: Some Reflections on Human Attitudes to Dogs, in *The Domestic Dog: Its Evolution, Behaviour and Interactions with People*, ed. J. Serpell, Cambridge University Press, Cambridge, UK, pp. 245–256.

Shermer, M. 2006. *Why Darwin Matters: The Case against Intelligent Design*, Henry Holt and Co., New York.

Shouse, B. 2003. Ecology: Conflict Over Cooperation, *Science 299*: 644–646.

Shugart, H. H. 2004. *How the Earthquake Bird Got its Name and Other Tales of an Unbalanced Nature,* Yale University Press, New Haven, CT.

Silko, L. M. 1977. *Ceremony*, Viking Penguin Books, New York.

Silko, L. M. 1981. *Storyteller*, Arcade, New York.

Silko, L. M. 1996. *Yellow Woman and a Beauty of the Spirit*, Simon and Shuster, New York.

Slipperjack, R. 1992. *Silent Words*, Fifth House, Saskatoon Canada.

Smith, A. M. 1993. *Shoshone Tales*, University of Utah Press, Salt Lake City, UT.

Smith, N. 1984. *Uneven Development: Nature, Capital, and the Production of Space*, Oxford University Press, New York.

Smith, N. 1996. The Production of Nature, in *FutureNatural: Nature, Science, Culture*, ed. G. Robertson, Routledge Press, New York, pp. 35–54.

Standing Bear, L. 1978. *Land of the Spotted Eagle*, University of Nebraska Press, Lincoln, NE.

Stearns, S. C. 1976. Life History Tactics: A Review of the Ideas, *Quarterly Review of Biology 51*: 3–47.

Stearns, S. C. 1992. *The Evolution of Life Histories*, Oxford University Press, New York.

Stephens, D. W. and J. R. Krebs. 1986. *Foraging Theory*, Princeton University Press, Princeton, NJ.

Stephenson, R. O. 1982. Nunamiut Eskimos, Wildlife Biologists, and Wolves, in *Wolves of the World,* eds. F. H. Harrington and P. C. Pacquet, Noyes, Park Ridge, NJ, pp. 434–439.
Stevens, S. 1997. Consultation, Co-management, and Conflict in Sagarmatha (Mt. Everest) National Park in Nepal, in *Conservation through Cultural Survival: Indigenous Peoples and Protected Areas*, ed. S. Stevens, Island Press, Washington, DC, pp. 63–98.
Stone, C. D. 1974. *Should Trees Have Standing? Towards Legal Rights for Natural Objects*, William Kaufman, Los Altos, CA.
Storm, H. 1972. *Seven Arrows*, Ballantine Books, New York.
Strum, S. C. and L. M. Fedigan. 2000. *Primate Encounters: Models of Science, Gender and Society*, University of Chicago Press. Chicago.
Stuart, A. J., P. A. Kosnitsev, T. F. G. Higham, and A. M. Lister. 2004. Pleistocene to Holocene Extinction Dynamics in Giant Deer and Wooly Mammoth, *Nature 431*: 664–669.
Suzuki, D., and P. Knudtson. 1992. *Wisdom of the Elders: Sacred Native Stories of Nature*, Bantam Books, New York.
Swann, J. A., ed. 1991. *The Power of Place: Sacred Ground in Natural and Human Environments,* Quest Books, Wheaton, IL.
Tanner, A. 1979. *Bringing Home Animals*, C. Hurst, London.
Tanner, A., R. McGrath, and C. Brice-Bennett. 2002. Traditional Innu Beliefs, in *"Spirituality" Among the Inuit and Innu of Labrador. Background Report for the Environmental Impact Assessment of the Voisey's Bay Mine and Mill Project.* http://www.innu.ca/tanner1.html.
Taylor, P. 1986. *Respect for Nature*, Princeton University Press, Princeton, NJ.
Taylor, P. 1992. The Ethics of Respect for Nature, in *The Animal Rights/Environmental Ethics Debate: The Environmental Perspective*, ed. E. C. Hargrove, State University of New York Press, Albany, NY, pp. 95–120.
Thomas, D. H. 2000. *Skull Wars: Kennewick Man, Archaeology and the Battle for Native American Identity*, Basic Books, New York.
Thomas, K. 1996. *Man and the Natural World: Changing Attitudes in England 1500–1800*, Oxford University Press, New York.
Thornton, A., and K. McAuliffe. 2006. Teaching in Wild Meerkats, *Science 313*: 227–229.
Tinbergen, N. 1953. *The Herring Gulls' World,* Basic Books, London.
Tinker, G. 2004. The Stones Shall Cry Out: Consciousness, Rocks, and Indians, *Wicazo Sa Review 19(2)*: 105–125.
Treat, J. 1999. Introduction to *For This Land: Writings on Religion in America*, by Vine Deloria Jr., Routledge Press, New York, pp. 1–18.
Trivers, R. L. 1972. Parental Investment and Sexual Selection, in *Sexual Selection and the Descent of Man, 1871–1971*, ed. B. Campbell, Aldine Press, Chicago, IL, pp. 136–279.
Trivers, R. L. 1974. Parent Offspring Conflicts, *American Zoologist 14*: 249–264.
Trosper, R. 2009. *Resilience, Reciprocity and Ecological Economics: Northwest Coast Sustainability*, Routledge, New York.
Trost, C. H. 1999. Black-billed Magpie, in *The Birds of North America*, eds. A. Poole and F. Gill, Philadelphia Academy of Natural Sciences, and American Ornithologist's Union, Washington, DC.
Turner, N. 2005. *The Earth's Blanket: Traditional Teachings for Sustainable Living*, University of Washington Press, Seattle, WA.
Venuto, Rochelle. 1998. Bone Game's Terminal Plots and Healing Stories, *Studies in American Indian Literature 10*: 23–42.

Vila, L., P. Savolainen, J. E. Maldonado, I. R. Amorim, J. E. Rice, R. L. Honeycutt, K. E. Crandall, J. Lundeberg, and R. K. Wayne. 1997. Multiple and Ancient Origins of the Domestic Dog, *Science 276*: 1687–1689.

Vizenor, G. 1981. *Earth Divers: Tribal Narratives on Mixed Descent*, University of Minnesota Press, Minneapolis, MN.

Vizenor, G. 1992. *Dead Voices: Natural Agonies in the New World*, University of Oklahoma Press, Norman, OK.

Vogel, G. 1998. DNA Suggests Cultural Traits affect Whales' Evolution, *Science 282*: 1616.

Waal, F. B. de. 1995. *Good-Natured: The Origins of Right and Wrong in Humans and Other Animals*, Harvard University Press, Cambridge, MA.

Waal, F. B. de. 2001. *The Ape and the Sushi Master,* Basic Books, New York.

Waal, F. B. de. 2009. *The Age of Empathy: Nature's Lessons for a Kinder Society*, Harmony Books, New York.

Waal, F. B. de, and P. L. Tyack. 2003. *Animal Social Complexity: Intelligence, Culture and Individualized Societies*, Harvard University Press, Cambridge, MA.

Wagamese, R. 1996. *Terrible Summer,* Warwick Publishing, Toronto.

Wallace, E., and E. A. Hoebel. 1948. *Comanches: Lords of the Southern Plains*, University of Oklahoma Press, Norman, OK.

Walters, C. J., M. Stocker, and G. C. Haber. 1981. Simulation and Optimization Models for a Wolf-Ungulate System, in *Dynamics of Large Mammal Populations*, eds. C. W. Fowler and T. D. Smith, Wiley-InterScience, New York, pp. 317–337.

Warren, D. M., L. J. Slikkerveer, and D. Brokensha, eds. 1995. *The Cultural Dimension of Development: Indigenous Knowledge Systems*, Intermediate Technology Publications, London.

Watanabe, K. 1994. Precultural Behavior of Japanese Macaques: Longitudinal Studies of the Kojima Troops, in *The Ethological Roots of Culture*, eds. R. A. Gardner, A. B. Chiarelli, B. T. Gardner, and F. X. Plooij, Kluwer Press, Berlin, pp. 81–94.

Webster's II New Riverside Dictionary. 1988. Houghton-Mifflin, Boston.

Weiner, J. 1995. *The Beak of the Finch*, Alfred A. Knopf, New York.

Welch, J. 1974. *Winter in the Blood*, Harper and Row, New York.

Welch, J. 1986. *Fools Crow*, Viking Press, New York.

West-Eberhard, M. J. 2003. *Developmental Plasticity and Evolution*, Oxford University Press, New York.

White, R. 2000. Review of Krech's *The Ecological Indian. New Republic*, June 24, 2000.

Whitehead, H. 1998. Cultural Selection and Genetic Diversity in Matrilineal Whales, *Science 282*: 1708–1711.

Whitehead, H. 2003a. Society and Culture in the Deep and Open Ocean: The Sperm Whale and other Cetaceans, in *Animal Social Complexity: Intelligence, Culture and Individualized Societies*, eds. F. B de Waal and P. L. Tyack, Harvard University Press, Cambridge, MA, pp. 444–464.

Whitehead, H. 2003b. *Sperm Whales: Social Evolution in the Ocean*, University of Chicago Press, Chicago, IL.

Whiten, A., J. Goodall, W. C. McGrew, T. Nishida, V. Reynolds, Y. Sugiyama, C. E. G. Tuten, R. W. Wrangham, and C. Boesch. 1999. Culture in Chimpanzees, *Nature 399*: 682–685.

Wilkinson, C. 2005. *Blood Struggle: The Rise of Modern Indian Nations*, W. W. Norton, New York.

Williams, G. W. 2002. Aboriginal Use of Fire: Are There Any Natural Plant Communities? in *Wilderness and Political Ecology: Aboriginal Influences and the*

Original State of Nature, eds. C. E. Kay and Williams, University of Utah Press, Salt Lake City, pp. 179–214.

Wilson, A. 1992. *The Culture of Nature: North American Landscape from Disney to the Exxon Valdez,* Blackwell, Cambridge, MA.

Wilson, E. O. 1975. *Sociobiology: The New Synthesis*, Harvard University Press, Cambridge, MA.

Wilson, E. O. 1978. *On Human Nature*, Harvard University Press, Cambridge, MA.

Wilson, E. O. 1994. *Naturalist*, Island Press, Washington, DC.

Wilson, E. O. 1998. *Consilience: The Unity of Knowledge*, Alfred Knopf, New York.

Wilson, G. L. 1929. Hidatsa Eagle Trapping, *Anthropological Papers of the American Museum of Natural History* 30: 99–245.

Wilson, K. 2005. MSY, No Net Loss and the Future of Fraser River Sockeye. *Proceedings From the World Summit on Salmon*, Chapter 10.

Wolff, L. 1994. *Inventing Eastern Europe: The Map of Civilization on the Mind of the Enlightenment*, Stanford University Press, Stanford, CA.

Wolpert, L. 1992. *The Unnatural Nature of Science*, Faber, London.

Worster, D. 1993. *The Wealth of Nature.* Oxford University Press, Oxford, UK.

Worster, D.1994. *Nature's Economy: A History of Ecological Ideas, 2nd Edition*, Cambridge University Press, Cambridge, UK.

Wright, J. P., and C. G. Jones. 2006. The Concept of Organisms as Ecosystem Engineers Ten Years on: Progress, Limitations, and Challenges, *Bioscience* 56: 203–209.

Wuerthner, G., ed. 2006. *Wildfire: A Century of Failed Forest Policy*, Island Press, Washington, DC.

Index

F

N

S

T

Y

Z

For Product Safety Concerns and Information please contact our EU representative GPSR@taylorandfrancis.com
Taylor & Francis Verlag GmbH, Kaufingerstraße 24, 80331 München, Germany

www.ingramcontent.com/pod-product-compliance
Ingram Content Group UK Ltd.
Pitfield, Milton Keynes, MK11 3LW, UK
UKHW022002180926
475701UK00003B/154
9780415517782